Efficiency, Performance and Robustness of Gas Turbines

Efficiency, Performance and Robustness of Gas Turbines

Editor

Albert Brose

scitus academics

Efficiency, Performance and Robustness of Gas Turbines
Edited by **Albert Brose**

ISBN: 978-1-68117-307-8
Library of Congress Control Number: 2016937615

www.scitusacademics.com
Box No. 4766, 616 Corporate Way,
Suite 2, Valley Cottage,
NY 10989

Printed in United States of America on Acid Free Paper

Preface

Gas turbine is the engine at the heart of the power generation plant. A gas turbine is an internal combustion engine type, also referred to as a combustion turbine. It has a rotating upstream compressor coupled with a downstream turbine and a burning chamber. The basic operation of the gas turbine is similar to the steam power plant except that the air is used in place of water. Fresh air flows through a compressor that increases the pressure. Energy is then added by sprinkling and igniting fuel in the air to produce a high temperature flow. This high-temperature high-pressure gas enters a turbine, reaching the exhaust pressure and producing a shaft during the process. The turbine shaft works to drive the compressor and other equipment, such as an electric generator, that can be connected to the shaft. A gas turbine converts to natural and other liquid fuels for mechanical energy. This energy then drives a generator producing electricity. Electricity moves to households and companies along power lines. The gas turbine heats a mixture of air and fuel to generate electricity at very high temperatures, causing the turbine blades to spin. The turbine drives a generator that converts power to electricity. The purpose of the gas turbine is to determine the design to maximize the desired energy. Gas turbines are used in aircraft, trains, ships, generators and tanks electricity.

This book covers a wide range of questions concerning the analysis and engineering of gas turbine applications. Diagnostic and condition monitoring systems for gas turbine engine defects, operating conditions for open gas turbines, noise reduction, gas turbine exhaust recovery, suitable materials and coatings, ultra-micro-gas turbines and gas turbine applications are discussed.

Table of Contents

1

CHAPTER

COMPARATIVE ASSESSMENT OF COMBINED-HEAT-AND-POWER PERFORMANCE OF SMALL-SCALE AERO-DERIVATIVE GAS TURBINE CYCLES

Barinyima Nkoi, Barinaadaa Thaddeus Lebele-Alawa

Mechanical Engineering Department, Faculty of Engineering, Rivers State University of Science and Technology, Port Harcourt, Nigeria

ABSTRACT

This paper considers comparative assessment of combined-heat-and-power (CHP) performance of three small-scale aero-derivative industrial gas turbine cycles in the petrochemical industry. The bulk of supposedly waste exhaust heat associated with gas turbine operation has necessitated the need for CHP application for greater fuel efficiency. This would render gas turbine cycles environ- mentally-friendly, and more economical. However, choosing a particular engine cycle option for small-scale CHP requires information about performances of CHP engine cycle options. The investigation encompasses comparative assessment of simple cycle (SC), recuperated (RC), and intercooled-recuperated (ICR) small-scale aero-derivative industrial gas turbines combined-heat-and- power (SS-ADIGT-CHP). Small-scale ADIGT engines of 1.567 MW derived from helicopter gas turbines are herein analysed in combined-heat-and-power (CHP) application. It was found that in this category of ADIGT engines, better CHP efficiency is exhibited by RC and ICR cycles than SC engine. The CHP efficiencies of RC, ICR, and SC small-scale ADIGT-CHP cycles were found to be 71%, 60%, and 56% respectively. Also, RC engine produces the highest heat recovery steam generator (HRSG) duty. The HRSG duties were found to be 3171.3 kW for RC, 2621.6 kW for ICR, and 3063.1 kW for SC. These outcomes would actually meet the objective of aiding informed preliminary choice of small-scale ADIGT engine cycle options for CHP application.

Keywords: Aero-Derivative Gas Turbines, Combined-Heat-and-Power, Heat Recovery Steam Generator, CHP Efficiency

1. INTRODUCTION

Gas turbine is a very satisfactory means of producing mechanical power. It is designed to be highly effective in producing aligned high thrust and power [1] [2] . In contemplating environmentally-friendly gas turbine cycles in the petrochemical industry identification is made of combined-heat-and-power (CHP) as one prominent application that would make gas turbine operation very pleasant to the environment in the aspects of reducing heat energy loss to the environment, and reducing global warming. It also enhances fuel efficiency. CHP simply defined is the simultaneous generation of mechanical power and heat energy in a single system from same fuel input [3] .

Some processes in the petrochemical industry occur at relatively moderate temperatures (below 600°C), and steam is generally the source of their heat energy supply. Such processes include the likes of refining and transformation of crude oil by separation, conversion, and purification carried out in refineries [4] , and steam cracking of heavier feedstock, polymerisation, and processing of aromatics occurring in petrochemical plants [5] . Steam could be generated by conventional boilers or heat recovery steam generators in CHP application. It is worth stating that combined-heat-and-power (CHP) generation of steam and power is presently a key energy saving, as well as environmentally-friendly technology in the petrochemical industry [5] .

The benefit of CHP is illustrated in Figure 1 where a CHP plant with a single 100 units fuel source yields about 75% overall cycle efficiency as against separate power plant and boiler source of an aggregate of 147 units of fuel yielding about 51% overall cycle efficiency [6] . In light of this, the performance of small-scale aero-derivative industrial gas turbines (SS-ADIGT) derived from helicopter gas turbines discussed by the author in ref [7] is herein analysed in CHP application.

The decision to use aero-derivative gas turbines is mainly based on economical and operational advantages. Gas turbine manufacturers have found that to reduce cost of designing and developing new gas turbines, a more effective approach is to develop high performance industrial gas turbines by modifying aircraft gas turbine engines [8] . Also, by introducing aero-derivative's removable gas generator, better flexibility is provided which in turn lead to reducing maintenance operation and enhancing gas turbine availability in industrial applications [9] . More so, implementing aero-derivative technology for industrial gas turbine has resulted in low maintenance downtime, good part-load efficiencies, and higher rate of return [10] . Besides, aero-derivative gas turbines can meet stringent NO_x control requirements because they are suitable for power augmentation by steam injection. For instance, the GE LM series industrial aero-derivative gas turbines are meeting NO_x requirements as low as 25 parts per million (ppm) using steam injection. Other merits of aero-derivative gas turbines include low weight-to-power ratio, compactness, and hence, lesser erection and startup time [11] [12] . Moreover, aero-de- rivative gas turbines are most suitable for highly efficient cogeneration plants, more flexible combined-cycle plants, and in mechanical drive applications for production and distribution of oil and gas [13] .

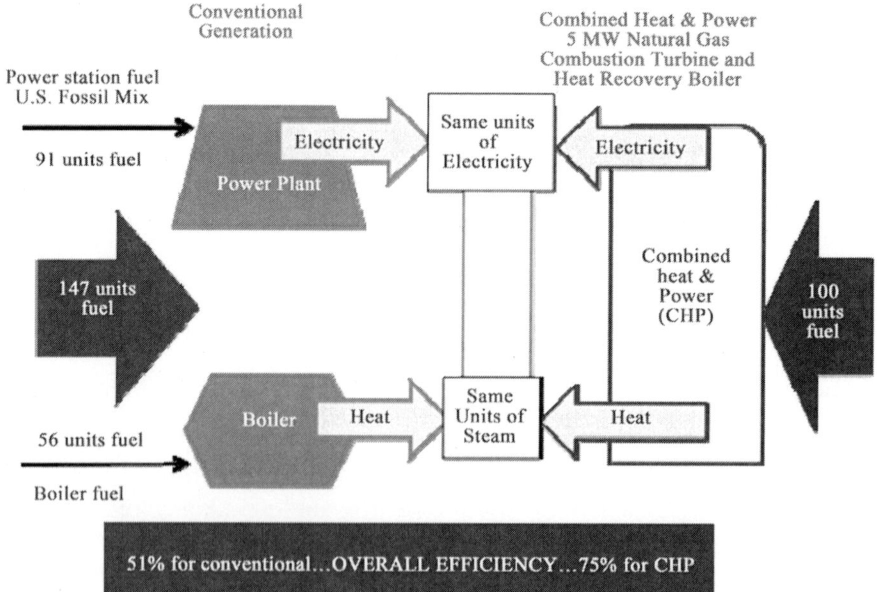

Figure 1. Energy saving benefit of CHP over traditional system [6] .

However, deciding on choice of small-scale ADIGT cycle option for CHP application poses some difficulty for engineers and decision-makers. Hence, the objective of this paper is to carry out performance comparison of simple cycle (SC), recuperated (RC), and intercooled-recuperated (ICR) small-scale ADIGT cycles in CHP, that would aid good and informed choice of turbine cycle option for the purpose of use in small-scale CHP application. The novelty of this research work is in the area of comparing the performances of SC, RC, and ICR aero- derivative gas turbine cycles in small-scale CHP. Previous works only considered CHP analysis of the simple engine cycle, and as such, considering CHP performance of advanced cycles (RC and ICR) actually presents a wider range of options for small-scale CHP engine cycle choices.

2. MATERIALS AND METHODS

2.1. CHP Modelling

CHP systems are either developed as "topping cycles" or "bottoming cycles" as illustrated in Figure 2 and Figure 3 respectively [14] . Topping cycles describe systems where there occur primary power generation and subsequent heat utilization, whereas bottoming cycles pertain to systems where heat is primarily generated in a process with subsequent utilisation for power generation [15] . In this work, topping cycle arrangement is adopted where power is primarily generated from gas turbine and heat recovery steam generator (HRSG) is designed to match for the purpose of process steam production. Performance parameters of the aero-derivative gas turbines discussed by the author in ref [7] are employed to determine the parameters of the HRSG.

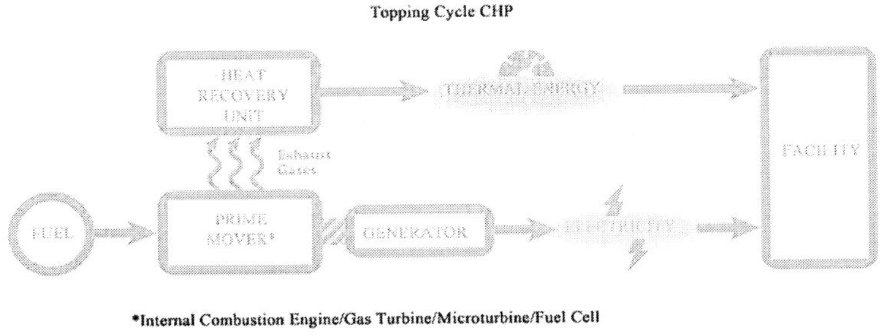

Figure 2. Topping cycle CHP (Source: [14]).

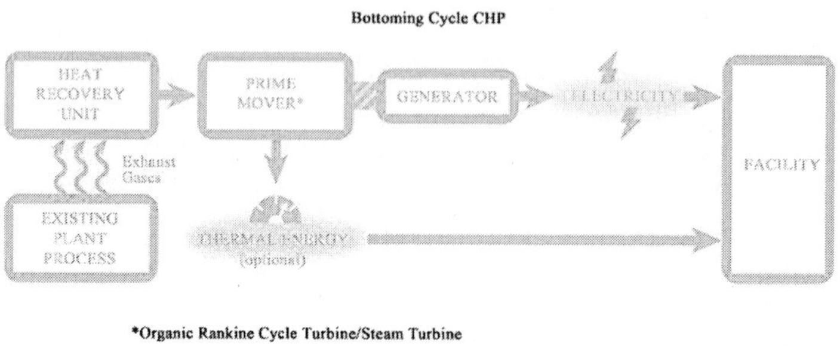

Figure 3. Bottoming cycle CHP (Source: [14]).

2.1.1. HRSG Performance

A set of heat exchangers that utilises the exhaust heat of a gas turbine to produce steam is referred to as heat recovery steam generator (HRSG). Three types of HRSG are identified, namely, unfired, supplementary fired, and exhaust fired. The most common and widely used HRSG is the unfired type because it is simple in design and cheap [16] . HRSG of the unfired type is considered in this research without considering the material dimension of the heat exchangers. It is pertinent to declare that only the thermodynamic performance in terms of temperature profile of exhaust gas, steam temperature and flow, and heat capacity, of the HRSG are being modelled in this research. Pinch and approach point technology is applied in modelling the HRSG performance, and with a single steam pressure mode of operation.

2.1.2. Pinch and Approach Points Technology

Approach point is the difference between the temperature of saturated steam and the temperature of water entering the evaporator, whereas pinch point is the difference between the gas temperature leaving the evaporator and the temperature of saturated steam [16] . Steam generation is directly affected by the pinch and approach points. Also affected is the exhaust gas and steam temperature profile. For the design case of an unfired HRSG, selection is usually made of the values of pinch and approach points; pinch point ranges from 10°C to 30°C whereas approach point ranges from 5°C to 15°C based on the sizes of evaporators that can be built and shipped economically, and to maximise heat transfer rate between exhaust gas and steam streams. Figure 4 illustrates pinch point, approach point, exhaust gas and steam temperature profiles of HRSG.

Using the notations in Figure 4 above the path 4-y-x-1 indicates gas turbine exhaust gas temperature profile whereas the path a-b-c-d-e indicates steam temperature profile. Pinch point $= T_x - T_c$; approach point $= T_c - T_b$; process a-b occurs in the economiser; c-d in the evaporator; and d-e in the super-heater.

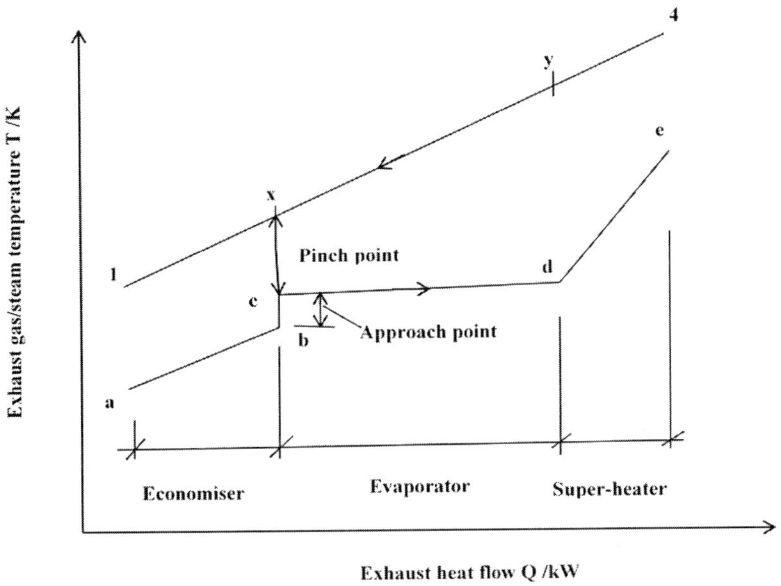

Figure 4. Single steam pressure HRSG exhaust gas/steam temperature profiles.

2.2. ADIGT-CHP Design Point Performance

To model the design point performance of a CHP plant is to match the parameters of HRSG with the design point of the gas turbine given particular consideration to desired steam flow or temperature and saturation pressure. In doing so, pinch and approach points are selected by the engineering judgement; and from gas turbine exhaust gas flow, the HRSG temperature profile, duty, and steam flow are established. Using pinch technology and thermodynamic properties of steam, the computation of CHP HRSG gas/steam temperature profile and steam flow is done as follows: Gas turbine exhaust gas temperature and mass flow are imported from gas turbine performance simulation while the HRSG pinch and steam saturation pressure (which fixes the steam saturation temperature—T_c) are selected by the engineering judgement. In this design the steam saturation pressure is 10bar. With the notations of Figure 4 the temperature of exhaust gas at pinch point (T_x) is given by Equation (1)

$$T_x = T_c + \text{pinch} = T_c + 15 \tag{1}$$

where $\text{Pinch} = 15$.

The superheated steam temperature (T_e) is chosen as required by the industrial process heat demand. The steam flow (w_s) is computed from total heat transfer in super-heater and evaporator using heat balance above pinch as defined by Equation (2)

$$Q_{4x} = Q_{evap} + Q_{super};\ Q_{4x} = w_g c_{pa}(0.99)(T_4 - T_x) = w_s\left[(h_e - h_c) + 0.02(h_d - h_c)\right]$$

$$\therefore w_s = \frac{w_g c_{pa}(0.99)(T_4 - T_x)}{(h_e - h_c) + 0.02(h_d - h_c)} \tag{2}$$

where 0.99 = heat loss factor,

0.02 = blow down factor,

$\dot{W}_g$ = exhaust gas flow,

$\dot{c}_{pa}$ = specific heat at constant pressure of air,

h_e = specific enthalpy of super-heated steam,

h_c = specific enthalpy of saturated water,

h_d = specific enthalpy of saturated steam,

T_4 = gas turbine exhaust temperature,

Q_{evap} = evaporator duty,

Q_{super} = super-heater duty.

Equation (3) defines the super-heater duty (Q_{super})

$$Q_{super} = w_s(h_e - h_d) \tag{3}$$

Gas temperature drop in the super-heater (ΔT_{4y}) is given by Equation (4)

$$\Delta T_{4y} = \frac{Q_{\text{super}}}{w_g c_{pa}(0.99)} \tag{4}$$

This implies that exhaust gas temperature to evaporator (T_y) is calculated using Equation (5)

$$T_y = T_4 - \Delta T_{4y} \tag{5}$$

Evaporator duty (Q_{evap}) is determined with the aid of Equation (6)

$$\text{Evaporator duty } Q_{\text{evap}} = w_s (h_d - h_c) \tag{6}$$

Similarly Equation (7) defines Economiser duty (Q_{econ})

$$\text{Economiser duty } Q_{\text{econ}} = w_s (1.02)(h_c - h_a) \tag{7}$$

Gas temperature drop in the economiser (ΔT_{x1}) is given by Equation (8)

$$\Delta T_{x1} = \frac{Q_{econ}}{w_g c_{pa}(0.99)} \tag{8}$$

This implies that exhaust gas exit temperature from the economiser (T_1) is calculated using Equation (9)

$$T_1 = T_x - \Delta T_{x14y} \tag{9}$$

Total HRSG duty (Q_{HRSG}) is computed by Equation (10)

$$\text{HRSG duty}(Q_{\text{HRSG}}) = Q_{\text{evap}} + Q_{\text{super}} + Q_{\text{econ}} \tag{10}$$

$$\text{Electrical efficiency} = \frac{P_E}{P_T} = \eta_E$$

The electrical efficiency could be assumed, such that

$$\text{Useful electric power generated } P_E = \eta_E + P_T$$

where P_T = gas turbine power.

Heat to power ratio of the CHP is given by Equation (11)

$$\text{Heat to power ratio} = \frac{Q_{\text{HRSG}}}{P_E} \tag{11}$$

Equations (1) to (11) are referred from [17]

Equation (12) is used to compute First Law CHP efficiency (η_1)

$$\text{First Law efficiency } \eta_1 = \frac{\dot{W}_E + \dot{W}_{ST}}{\dot{m}_f \cdot h_f} \tag{12}$$

where $\dot{W}_E$ = Electrical energy rate,

$\dot{W}_{ST}$ = Steam energy rate,

$\dot{m}_f$ = Fuel mass flow in combustor,

Δh_f = LHV = Low heating value of fuel.

Equation (13) is used to compute Second Law CHP efficiency (η_2)

$$\text{Second law efficiency } \eta_2 = \frac{\dot{W}_E + \dot{W}_{ST}}{\dot{m}_f \left[\Delta h_f - T_R \cdot \Delta S_f\right]} \tag{13}$$

The denominator of Equation (13) is the availability rate of the fuel consumed, where ΔS_f = Entropy released by fuel combustion. T_R = Temperature at exhaust. Equations (12) and (13) and referred from [18] [19] .

3. RESULTS AND DISCUSSIONS

3.1. Small-Scale-ADIGT-CHP Design Point Performance Analysis

The CHP design point simulation for the SS-ADIGT was done using TURBOMATCH (a gas turbine engine performance simulation code) [7] - [22] and the HRSG gas/steam temperature profiles are shown in Figure 5 while the CHP DP performance results are indicated in Table 1.

The technical performance of the simple, recuperated, and intercooled-recuperated SS-ADIGT cycles derived from helicopter engines have been analysed by the author in [7] . The superheated steam temperature for the

CHP was set within the range of steam temperatures obtainable in refinery and petrochemical plants which is about 100°C - 500°C.

3.2. Small-Scale ADIGT-CHP Off-Design Performance

The HRSG would normally not operate at the design point due to variations in the inlet gas conditions and steam parameters. The inlet gas conditions in turn would depend on gas turbine off-design variation in ambient conditions, firing temperature, altitude, power setting, etc. This makes the CHP plant exhibits varying outputs. The CHP off-design performance was simulated with TURBOMATCH engine off-design. The off-design performances of the SS-ADIGT-CHP with changing conditions of the engines are shown in Figures 6-10.

Table 1. SS-ADIGT-CHP design point performance results.

Parameter	Values for the SS-ADIGT engines		
	Simple cycle	Recuperated	ICR
Steam saturation temperature (K)	457	457	457
Pinch point	15	15	15
Approach point	8	8	8
Superheated steam temperature (K)	673	673	673
Steam mass flow (kg/s)	1.10	1.14	0.94
Economiser feed water temperature (K)	388	388	388
Super-heater duty (kW)	530.07	548.80	453.69
Evaporator duty (kW)	2198.23	2275.89	1881.45
Economiser duty (kW)	334.74	346.57	286.50
HRSG duty (kW)	3063.05	3171.26	2621.63
Gas turbine exhaust mass flow (kg/s)	5.65	5.64	5.64
Gas turbine exhaust temperature (K)	885	901	826
Gas temperature at evaporator exit (K)	806	819	758
Gas exit (stack) temperature (K)	422	420	429
Gas turbine power (kW)	1567	1567	1567
GT Thermal efficiency	**0.296**	**0.336**	**0.339**
Heat: power ratio	2.09	2.16	1.79
CHP efficiency	0.56	0.71	0.60

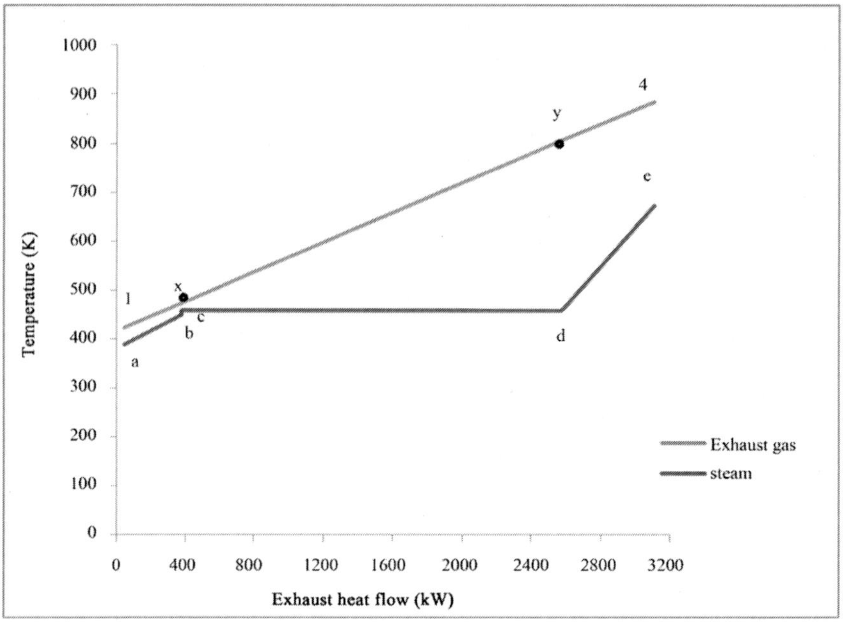

Figure 5. Single steam pressure HRSG temperature/heat profile for the SS-ADIGT-CHP.

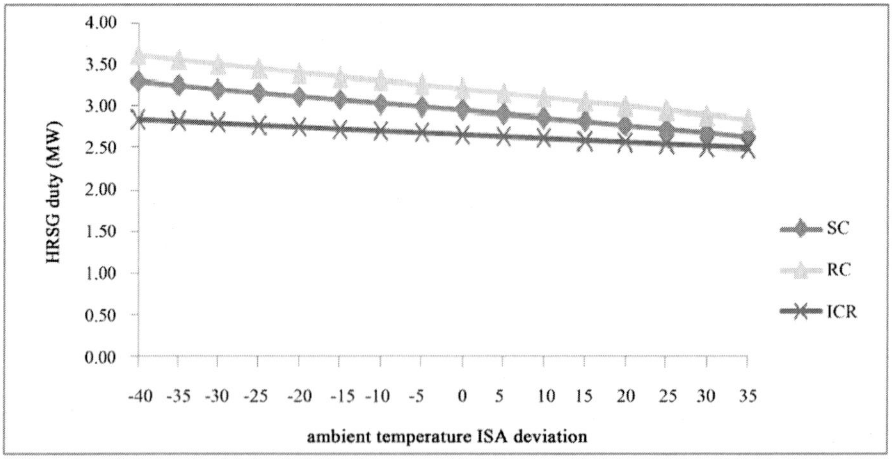

Figure 6. Effect of ambient temperature on HRSG duty.

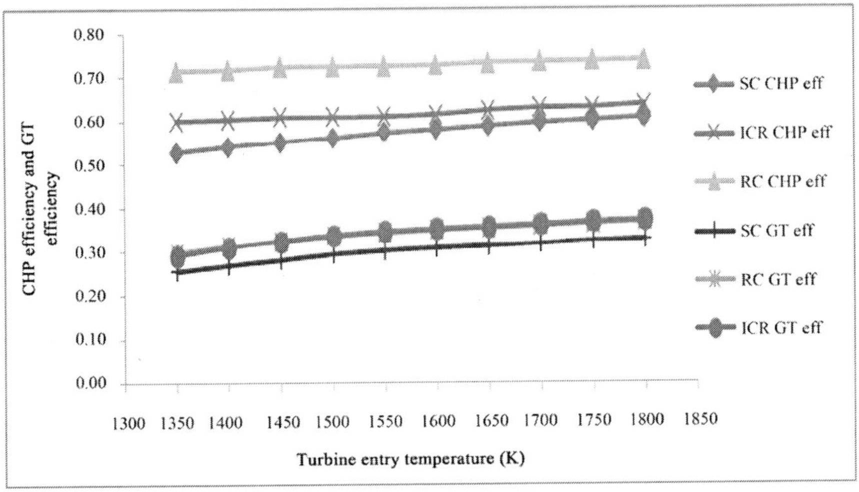

Figure 7. Variation of CHP and GT efficiencies with TET.

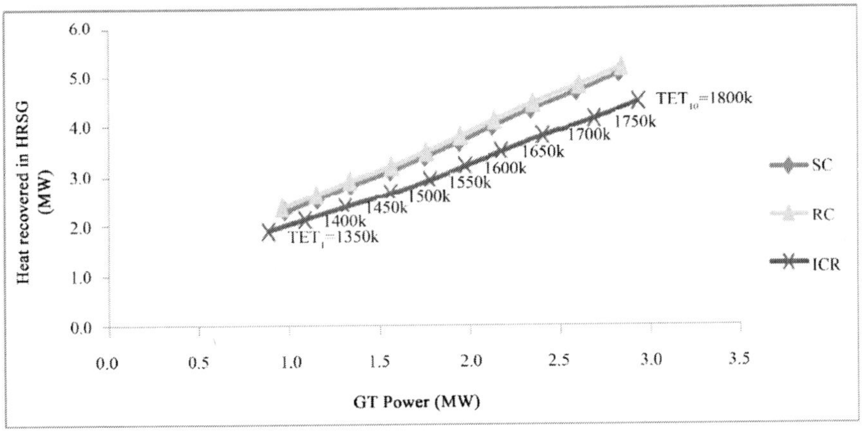

Figure 8. Variation of HRSG duty with GT power at increasing TET.

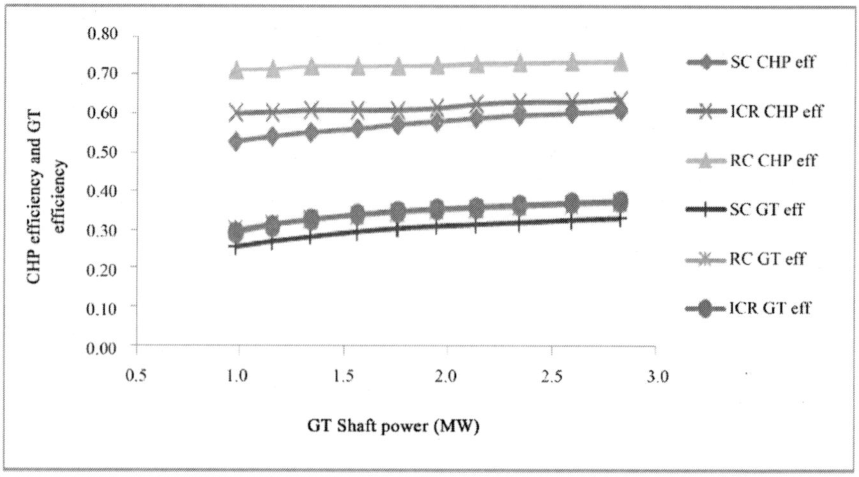

Figure 9. Variation of CHP and GT efficiencies with GT power at increasing TET.

At design and off-design conditions the RC and ICR ADIGT engines exhibit better CHP efficiency than the SC engine as shown in Figure 7, Figure 9, and Figure 10. As shown in Table 1 the CHP efficiencies of RC, ICR, and SC SS-ADIGT-CHP cycles were found to be 71%, 60%, and 56% respectively. These results compare favourably with values in the literature. For instance, a 40 kWe CHP plant located within the Queens Building at De Montfort University, was analysed to show an overall (CHP) efficiency of 77% [23] . CHP efficiency can be as high as 80% - 90%. For instance Tervola 0.5 MWe/1.13MWth CHP in Finland was found to exhibit an over- all (CHP) efficiency of 81.5% [24] . The CHP efficiencies are observed to increase with increases in turbine en- try temperature (TET), gas turbine (GT) power, ambient temperature, and HRSG duty. The percentage increases in CHP efficiencies of RC and ICR over SC at design-point (DP) are 16.5% and 3.8% respectively. This superior performance is due to the lower heat input from burning less fuel in the advanced cycle engines. Looking at Figure 7 and Figure 9, the curve "RC GT eff" appears to coincide with that of the "ICR GT eff", whereas actually there is some slight difference between the RC GT efficiency and ICR GT efficiency. This small difference is shown clearly at design point in Table 1 where ICR GT efficiency is slightly higher than RC GT efficiency by about 0.003.

On the other hand, the SC engine produces more HRSG duty than the ICR cycle as shown in Figure 6 and Figure 8, due to lower exhaust gas temperature and steam rate of the ICR cycle. The highest HRSG duty is produced by the RC engine because of its higher exhaust gas temperature and steam rate.

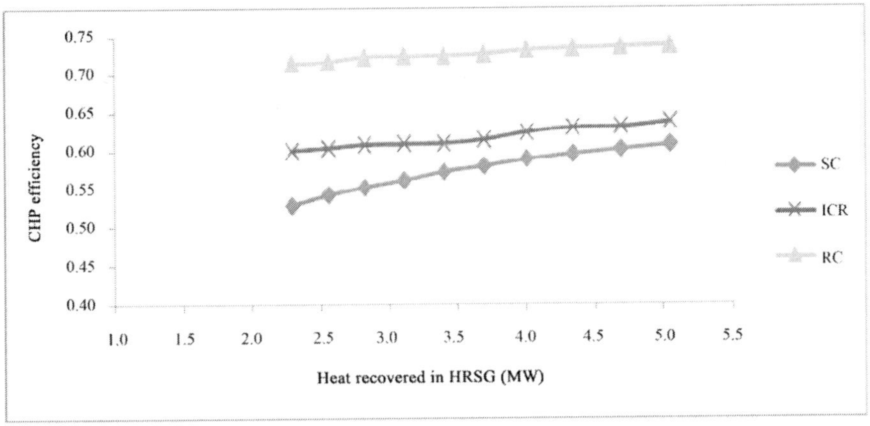

Figure 10. Variation of CHP efficiency with HRSG duty at increasing TET.

Besides, as plotted in Figure 8, the heat-to-power ratio is observed to be least in the ICR cycle than SC and RC and highest in the RC cycle at both DP and off-design (OD) conditions. In essence, this means that given a range of power output at any condition of TET and ambience, the RC cycle would generate more steam than the SC and ICR, while ICR would generate the least steam flow. This is so because GT exhaust gas temperature is highest in the RC cycle and exits the HRSG at the least temperature in the RC cycle. This creates huge drop in temperature of exhaust gas from exit of GT to exit of HRSG in the RC cycle. This exhaust gas temperature drop is smaller in SC cycle and least in ICR cycle. Hence, huge amount of heat is extracted from exhaust gas in RC cycle than SC and ICR at same conditions. At design point, the HRSG duties were found to be 3171.3 kW for RC, 2621.6 kW for ICR, and 3063.1 kW for SC as shown in Table 1.

4. CONCLUSION

The foregoing analysis of technical performances of small-scale aero-derivative industrial gas turbine-CHP cycles has led to the conclusions that for small-scale ADIGT-CHP, better CHP efficiency is exhibited by RC and ICR cycles than Simple engine cycle. Also, it was found that the RC engine produces the highest HRSG duty. Therefore, it could be said that performance comparison of simple, recuperated, and intercooled-recuperated small-scale ADIGT cycles in CHP has been achieved. This sort of analysis would actually aid concerned engineers, product developers, and other key decision-makers to logically make good and informed choice of small-scale ADIGT engine cycle option for the purpose of use in CHP application.

ACKNOWLEDGEMENTS

Very essentially, the authors would want to thank Professor Pericles Pilidis and Dr. Theoklis Nikolaidis of the Department of Power and Propulsion of Cranfield University, United Kingdom, for their invariable contributions to this research.

REFERENCES

1. Lebele-Alawa, B.T. and Jo-Appah, V. (2015) Thermodynamic Performance Analysis of a Gas Turbine in an Equatorial Rainforest Environment. Journal of Power and Energy Engineering, 3, 11-23.
2. Lebele-Alawa, B.T. (2010) Axial Thrust Responses to a Gas Turbine's Rotor-Blade Distortions. Journal of Engineering Physics and Thermophysics, 83, 991-994.
3. International Energy Agency (2007) Tracking Industrial Energy Efficiency and CO2 Emissions.
4. Exxonmobil (2011) Fawley Refinery and Petrochemical Plant. www.exxonmobil.co.uk/UK-English/files/Fawley_2011.pdf
5. Gielen, D.J., Vos, D. and Van Dril, A.W.N. (1996) The Petrochemical Industry and Its Energy Use: Prospect for the Dutch Energy Intensive Industry. Energy Research Center of the Netherland (ECN), ECN-C-96-029. April 1996, 25-45.
6. US EPA (2013) Combined Heat and Power Partnership: Efficiency Benefit.

7. Nkoi, B., Pilidis, P. and Nikolaidis, T. (2013) Performance of Small-Scale Aero-Derivative Industrial Gas Turbines Derived from Helicopter Engines. Journal of Propulsion and Power Research, 2, 243-253.

8. Bhargava, R., Blanchi, M., Peretto, A. and Spina, P. R. (2004) A Feasibility Study of Existing Gas Turbines for Recuperated, Intercooled, and Reheat Cycle. Journal of Engineering for Gas Turbines and Power, 126, 531-544.

9. Najjar, Y.S.H. (2000) Gas Turbines Cogeneration Systems: A Review of Some Novel Cycles. Applied Thermal Engineering, 20, 179-197.

10. Yang, W. (1997) Reduction of Specific Fuel Consumption in Gas Turbine Power Plants. Energy Conversion and Management, 38, 1219-1224.

11. Keller, S.C. and Studniarz, J.J. (1987) Aero-Derivative Gas Turbines Can Meet Stringent NOx Control Requirements. Proceedings from the 9th Annual Industrial Energy Technology Conference, Houston, 16-18 September 1987, 253-260.

12. Roy, G.K. (2012) Selecting Heavy-Duty or Aero-Derivative Gas Turbines. Hydrocarbon Processing, 75, 57.

13. Doom, T.R. (2013) Aero-Derivative Gas Turbines. Case Studies on the Government's Role in Energy Technology Innovation. American Energy Innovation Council.

14. Center for Sustainable Energy (2014) Combined Heat and Power.

15. Bhatt, S.M. (2001) Mapping of General Combined-Heat-and-Power System. Energy Conversion and Management, 42, 115-124. www.elsevier.com/locate/enconman

16. Ganapathy, V. (1996) Heat Recovery Steam Generator: Understanding the Basics. Chemical Engineering Process, 32-45.

17. Ganapathy, V. (1990) Heat Transfer—Simplify Heat Recovery Steam Generator Evaluation. Hydrocarbon Processing, March 1990, 77-82.

18. Koratianitis, T. (2012) DEN107 Thermodynamics Laboratory Exercise [Course Lecture Note]. Fundamentals of General Thermodynamics. Queen Mary University of London, London.

19. Koratianitis, T., Grantstrom, J., Wassingbo, P. and Massardo, A.F. (2005) Parametric Performance of Combined-Co-generation Power Plant with Various Power and Efficiency Enhancements. Journal of

Engineering for Gas Turbines and Power. Transactions of the ASME, 127, 65-72.

20. Pachidis, V.A. (2008) Gas Turbine Performance Simulation [Course Lecture Note]. Gas Turbine Simulation and Diagnostics. Cranfield University, Cranfield.

21. Palmer, J. (1999) The Turbomatch Scheme for Aero/Industrial Gas Turbine Engine Design Point/Off Design Performance Calculation [Course Lecture Note]. Gas Turbine Simulation and Diagnostics. Cranfield University, Cranfield.

22. Nkoi, B., Pilidis, P. and Nikolaidis, T. (2013) Performance Assessment of Simple and Modified Cycle Turboshaft Gas Turbines. Journal of Propulsion and Power Research, 2, 96-106.

23. Smith, M.A., Few, P.C. and Twidell, J.W. (1995) Technical and Operational Performance of a Small-Scale Combined-Heat-and-Power (CHP) Plant. Energy, 20, 1205-1214.

24. 24. Kirjavainen, M., Sipila, K., Alakangas, E., Savola, T. and Salomon, M. (2004) Small-Scale Biomass CHP Technologies Situation in Finland, Denmark and Sweden. OPET Report 12, European Commission Directorate-General for Energy and Transport. OPET CHP/DHC Cluster. Espoo, April 2004, 27.

2

CHAPTER

IMPROVING THE TRANSIENT PERFORMANCE OF THE GAS TURBINE BY STEAM INJECTION DURING FREQUENCY DIPS

Saeed Bahrami [1], Ali Ghaffari [1] and Marcus Thern [2,*]

[1] *Department of Mechanical Engineering, K.N. Toosi University of Technology, Pardis Street, Vanak Square, Tehran 19991 43344, Iran*

[2] *Division of Thermal Power Engineering, Department of Energy Sciences, Lund University, Ole Römers väg 1, Lund SE-221 00, Sweden*

ABSTRACT

Single-shaft gas turbines are sensitive to frequency changes which might affect the grid stability during large frequency drops. This paper presents a new control system that uses steam injection as an auxiliary input to improve the transient performance of the gas turbine during frequency drops. Steam injection is beneficial because it reduces the peak temperature in the combustion chamber and augments the output power by increasing the mass flow through the turbine. The use of this auxiliary input is based on the event-based control approach. It means that during the frequency drop, the controller exploits the steam injection to help the main control loop recover the frequency and when the frequency reaches its predefined value, the system will return to its normal operation. The performance of the proposed control algorithm is investigated under different scenarios and the results show that the application of steam injection improves the performance of the regular control algorithm significantly, especially near full load condition.

Keywords: single-shaft gas turbine; combined-cycle; frequency control; steam injection

1. INTRODUCTION

Gas turbines have, during the last decades, become more common as a power generating engine due to their flexibility and low specific investment costs. Development of gas turbines for power generation has gone hand in hand with the development of new gas turbines for use in aircraft engines. This increase in the proportion of gas turbine and combined-cycle units in power generation grids will affect the grid characteristics because of the gas turbine performance. Gas turbines will behave differently than other prime movers, e.g., hydroelectric units, when a frequency drop occurs in the system [1]. The difference occurs because the dynamic behavior of the gas turbine depends on its rotational speed, which is proportional to the grid frequency. A drop in the grid frequency lowers the rotational speed of the gas turbine which affects the performance of the gas turbine and this can lead to unit instability during large frequency dips [2]. This issue is more crucial when the gas turbine is used in an isolated grid or in grids with large renewable energy capacity and smart grids. To have a good frequency regulation in a power grid, there should be a balance between production and demand. Unfortunately, the amount of energy that is supplied by renewable energy resources such as wind and solar energy to the grid is not controllable. In other words, unlike conventional sources of electric power, these renewable sources are not "dispatchable" [3]. Consequently, in smart grid and intermittent renewable energy dominated power grids, the fossil fuel units are usually responsible for grid frequency regulation and these units should be able to compensate the variation in power production of renewable energy units. Therefore, the fossil fuel in these grids should be able to respond faster and their performance should be more tolerant to frequency changes.

In modern combined-cycle power plants, the steam part is usually working on sliding pressure for higher efficiency, which means that there is not a lot of energy stored in the steam cycle that can be used for fast transient response. To respond to a frequency drop, the gas turbine must provide the energy to the steam cycle but it takes a while (in the order of 20 s) before the increase in fuel flow into the gas turbine combustion chamber affects the steam production. Therefore, in modern combined-cycle power plants, the transient performance of the unit depends entirely on gas turbine performance. Consequently, it would be reasonable to neglect the transient response of the steam part in analyzing the unit performance during frequency drops [4].

Single-shaft gas turbines are usually used in combined-cycle power plants and large power generation units because of their simplicity, lower cost, no gearbox, better heat recovery and, as a result, higher efficiency. In single-shaft gas turbines, compressor, turbine and generator are located on an identical shaft. This configuration increases the generator's inertia and consequently, it would minimize the speed reduction, if there is a step increase in engine load. Also, they are faster than multi shaft engines because there is no lag while a gas generator accelerates to increase the exhaust gas power to accelerate the free power turbine [5]. On the other hand, reduction in grid frequency would decrease the compressor speed and as a result the air mass flow rate will also decrease during frequency dips. This reduction in air mass flow rate will put limitation on the amount of engine overfueling to protect the engine from overheating. Engine overfueling is required to accelerate the turbine to recover the frequency and this limitation may lead to unit instability in large frequency dips or near full load condition [6]. This phenomenon is an inherent problem of the single-shaft gas turbines and it has nothing to do with the controller design.

Another way of combining steam and gas turbine is the steam-injected gas turbine (STIG). In the STIG, the hot exhaust gases from the gas turbine heats water to produce steam in a heat recovery steam generator. The steam is then led back to the gas turbine expander. This would result in higher efficiency, higher output power and lower emissions [7]. There have been many different ways of injecting steam. The steam is injected to the combustor liners for power augmentation but it would be injected into or around the fuel nozzles to reduce the NO_x levels [8,9,10,11]. Today, the STIG cycle is commercially available around the world.

The objective of this paper is to study how steam injection into the combustion chamber can improve the transient performance of gas turbines during frequency drops. This paper presents a control algorithm based on an event-based control approach. This means that if a frequency drop occurs, the controller uses steam injection to help the main control loop to recover the frequency. When the frequency reaches its predefined value, the system then continues with its regular control algorithm. The performance of the proposed control algorithm is investigated under different scenarios during the frequency drops in island mode, which can be known as the extreme case in frequency regulation. Also, the effect of application of the new control algorithm on important gas turbine variable such as surge margin is considered in this paper.

This paper is organized in six sections. First, the regular control logic of the gas turbine and it shortcomings are discussed. Section 3 presents the proposed solution and detail of new control system. It follows by a discussion about the effect of the steam injection on the surge margin. The simulation results and conclusions are presented in Section 5 and Section 6.

2. REGULAR GAS TURBINE CONTROL SYSTEM AND ITS SHORTCOMINGS

Gas turbines are usually equipped with a complicated control system and include several different loops such as governor, temperature controller, acceleration controller, *etc*. The governor and temperature controller loop are the main control loops that affect the performance of the gas turbine during frequency dips. For this reason, only these two control loops will be considered here. The governor is known as the main control loop and it controls the gas turbine during normal operation. It adjusts the fuel flow into the combustion chamber so that the power output and speed will be equal to whatever the operator requires. Moreover, in combined-cycle power plants, the gas turbine exhaust temperature should be kept as high as possible for better heat recovery because it results in a higher overall efficiency. In order to maximize exhaust temperature, the temperature controller closes the inlet guide vanes (IGVs) to reduce the air mass flow rate through the engine in part load. The temperature controller also maintains the temperature of gas turbine components within the design range for safe and reliable operation. Consequently, the temperature controller can also reduce the fuel flow into the combustion chamber when the firing temperature reaches to its limit. The overheating protection loop is only activated in the fast engine loading or when the engine is working near its full load condition. The detail of the governor and temperature controller for a sample gas turbine is given in Appendix A.

Figure 1 shows how the governor and the temperature control loop are connected to other components. The fuel flow calculated by the governor and the temperature controller are fed into a low vale selector (min) and their minimum is applied to the engine. This control logic provides a bumpless transfer between different control loops and this also means that at each moment only one controller is active. Although this control structure is very efficient and convenient under most of the operating conditions, it has two inherent problems in handling large frequency dips.

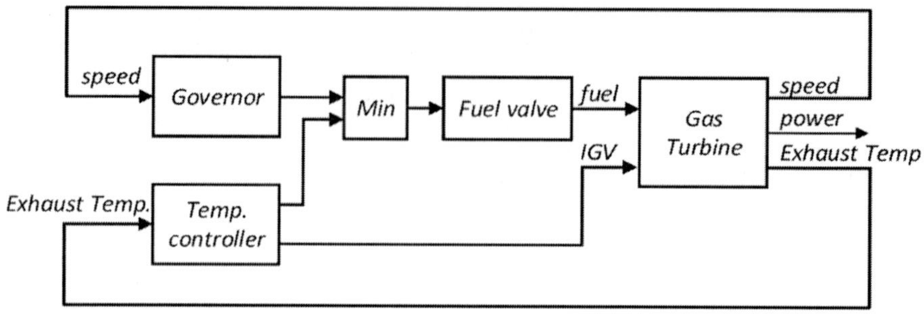

Figure 1. Gas turbine regular control configuration in island mode.

2.1. Gas Turbine Cannot Recover Frequency Dips near Full Load Condition

In single-shaft gas turbines, the compressor and generator are connected together. Therefore, during frequency dips, the compressor will rotate with a speed that is lower than its design speed. This reduction in rotational speed will lead to a decrease in air mass flow rate through the engine and result in a higher firing temperature. This is a transient effect and as soon as the gas turbine recovers the frequency, the air mass flow will be recovered as well. Additionally, the frequency drop is the consequence of higher power demand in the grid, so the control algorithm should apply more fuel to the combustion chamber to meet the new demand. The control algorithm will also need to provide some overfueling to accelerate the turbine so that it can recover the frequency at the same time. The fuel, which is related to new power demand, is required even after the frequency recovery but the overfueling which is related to turbine acceleration is a transient effect and it is not necessary after the speed returns to its nominal value. More fuel into the combustion chamber means a higher firing temperature, so the control algorithm tries to open the IGVs further and increase the air mass flow through the turbine to protect the engine from overheating. The IGV opening should be able to compensate all of the three effects, the air mass flow decrease because of reduced speed, overfueling for turbine acceleration and more fuel for the new power demand. The first two effects are transient and they are not necessary after the frequency is recovered but the IGV opening for the new power demand should be maintained even after frequency recovery. The required IGV opening is shown in Figure 2 schematically.

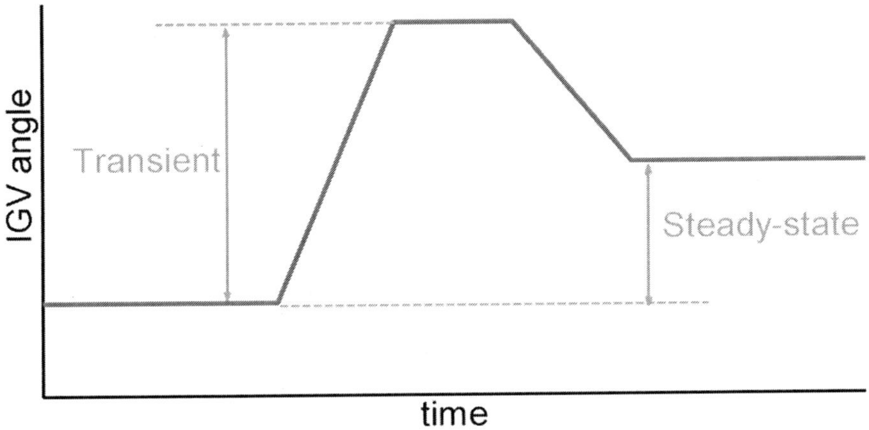

Figure 2. Schematic IGVs opening during frequency dips.

There is no problem with current control logic until the turbine is near full load condition. Near full load, the IGVs are almost fully open and this means that the controller doesn't have the needed reserve in IGVs which is required to provide the necessary air flow through the engine. This means that the firing temperature reaches its limit and the temperature controller will therefore reduce the fuel flow into the combustion chamber and will prevent the turbine from being loaded further. This leads to a continuous decrease in speed and creates a negative spiral that renders the gas turbine unstable. This situation is not acceptable when the IGVs have the reserve for new load demand but not enough reserve for the turbine acceleration and compensating the reduction in air mass flow because of speed reduction, since they are just needed for short period of time.

2.2. Slow Response of the Gas Turbine

The air flow dynamic is slower than the fuel flow dynamic. Fuel can be added directly into the combustion chamber, which means that it affects the output power and firing temperature quickly. On the other hand, to increase the air flow, the IGVs should be opened first and then the air accelerates and passes through the compressor stages to reach to the combustion chamber. Therefore, it affects the power output and firing temperature with a lag. This is due to the air inertia and IGV's actuator speed which is usually slower in comparison to a fuel valve. The time it takes for the gas turbine to respond to a frequency dip is therefore dominated by the airflow dynamic. This is important because a fast response to a new power demand is equivalent to a smaller decrease in speed. The smaller speed reduction means less reduction in airflow and less overfueling for turbine acceleration. Consequently, the gas turbine governor can recover the frequency easier without reaching the safety temperature limits.

Although some solutions such as using fast IGVs actuator [12], using the IGVs to load the engine and fuel to control the firing temperature [13] and application of new control algorithms [14] are suggested to solve these problems, but they are just useful for second problem and they have nothing to do with the first problem.

3. PROPOSED SOLUTION

The proposed solution is to use steam injection into the combustion chamber as an auxiliary input to cover the shortcoming of current control algorithm during large frequency drops and frequency dips near full load condition. The application of steam injection is beneficial for several reasons. First of all, it augments the output power of the gas turbine by increasing the mass flow and heat capacity of working fluid through the turbine [5]. Also, because the steam is injected directly into the combustion chamber, it affects the output power very fast. Moreover, the effect of steam injection, like fuel flow, doesn't depend on the speed of the gas turbine because it doesn't need to wait for the increase in airflow. Additionally, steam injection reduces the firing temperature of the gas turbine and therefore, the governor can overfuel the engine without reaching safety temperature limits. It also reduces the gas turbine emissions [15,16,17] and increases the service life of the engine components by decreasing the peak temperature in the combustion chamber during the transient maneuvers. On the other hand, it has its own drawbacks; steam injection might reduce the compressor surge margin [18] and also, it needs treated and purified water; otherwise it will affect the blade life [7].

Application of this auxiliary input will be with an event-based control approach. It means, only if the frequency drop passes a certain limit, the controller will use steam injection to help the governor to recover the frequency and when the speed reaches its predefined value, the controller will stop the steam injection. It should be noted that, in combined-cycle power plants or in combined heat and power (CHP) units, the required steam is already at hand and new control system can be implemented easily with a small modification in original system. Also, for small amount of steam injection, there is no need for gas turbine or combustion chamber modification [19].

To implement the proposed solution, a new loop is added for steam injection to the gas turbine original control system. This loop which is shown in Figure 3 includes following elements.

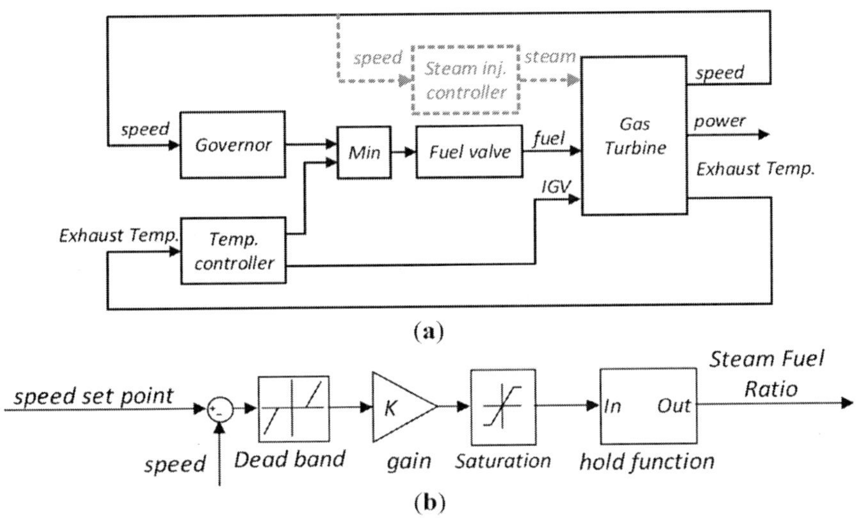

Figure 3. (**a**) New control loop configuration; and (**b**) steam injection controller details.

3.1. Dead Band

The gas turbine original control system can recover small frequency dips, so it is not necessary to use steam injection. Also, as stated in the previous section good water quality is required for steam injection. Producing such a high quality steam is quite expensive and using steam injection for small frequency dips will be a waste of resources.

3.2. Saturation

The controller should act as fast as possible to have successful frequency recovery, otherwise the situation will become worse and more difficult to handle. Therefore, a high gain controller is used in this loop. But large amount of steam injection might push the engine into the unstable working regions, so to restrict the amount of steam that can be injected into the gas turbine a saturation element is added to this loop.

3.3. Hold Function

With regular control algorithm, the amount of injected steam will be proportional to the gas turbine speed deviation from its nominal value. So the controller will cut off the steam injection as soon as it recovers the frequency. If, however, the steam injection continues for a while and decreases with a slow rate, it can provide the required time for the governor and IGVs to move to the new operating condition more smoothly. This would reduce the temperature gradients in the engine component and as a result, extend the service time of these components. So the amount of injected steam should be increased as fast as possible and reduces in a slow rate after frequency recovery. The hold function which is shown in Figure 4a is designed for this purpose. The time that controller should continue steam injection after frequency recovery can be adjusted easily by tuning the lower limit of the saturation block. The behavior of the developed hold function to a sample input is shown in Figure 4b for different hold times.

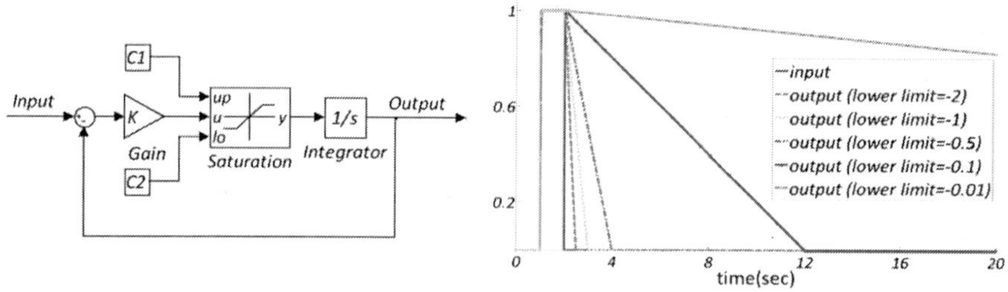

Figure 4. (**a**) Hold function; and (**b**) the response of the hold function to a sample function (gain = 100, upper limit = 1000).

4. SURGE MARGIN

Compressor surge is associated with a sudden drop in delivery pressure, and violent aerodynamic pulsation which is transmitted throughout the whole machine [20] and it should be avoided to protect the engine from damage. It happens when the delivery pressure of the compressor is lower than the downstream pressure. In single-shaft engines used for power generation, the combustor pressure is proportional to the engine load, so steam injection and over fueling which are increasing the output power will lead to pressure rise in the combustion chamber as well. Consequently, large amount of steam injection or overfueling may cause the gas turbine compressor to surge.

In the compressor characteristic map, the surge region is specified with a surge line (Figure 5). The compressor should operate within a safe distance from the surge line because of different reasons such as production tolerances, engine aging, and transient allowance, *etc.* The parameter for quantifying this distance is called surge margin and it denotes the relative distance of working point from the surge line. The surge margin is defined as:

$$SM(\%) = \frac{Pr_{surg} - Pr_{working}}{Pr_{working}} \times 100 \qquad (1)$$

where *Pr* stands for compressor pressure ratio. The normal surge margin which is considered for transient maneuvers is about 10%–15% [5]. The important point that should be noted here is that, the surge margin depends on the compressor speed and it decreases by speed reduction [4].

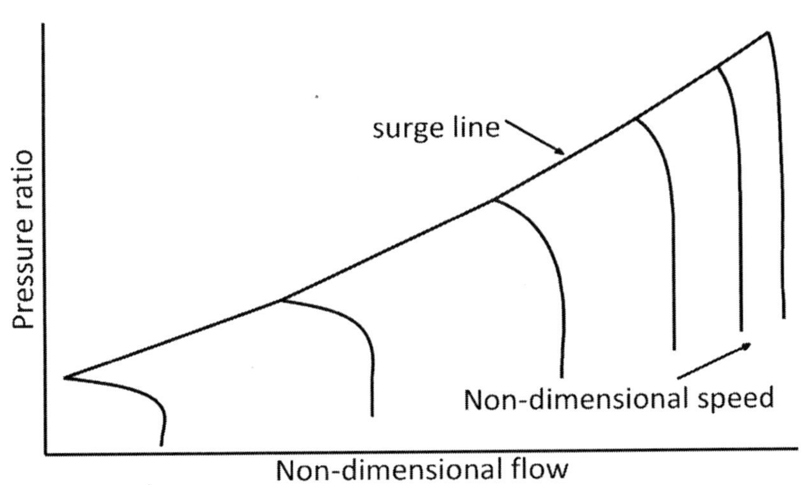

Figure 5. Schematic compressor map.

To compare the effect of steam injection verses overfueling on the compressor surge margin reduction, some simulations has been done and the results are summarized in Table 1. In these simulations, the output power of the gas turbine is kept constant with different amount of steam injection. As it can be seen, by increasing the amount of injected steam, the compressor pressure ratio will be increased slightly and at same time the firing temperature will be lower. So for the same output power, steam injection will reduce the surge margin more than fuel.

Table 1. Dry engine *vs.* steam injected engine parameters.

Engine load [MW]	**161**	**161**	**161**
steam-fuel ratio(SFR)	0	1	2
pressure ratio	11.66	11.78	11.91
ırbine inlet temperature (TIT) [k]	1333	1311	1305

5. SIMULATION RESULTS

A 160 MW heavy-duty gas turbine was studied in order to investigate the performance of the proposed control algorithm. The complete physical model for a dry engine is developed based on the heat and mass balance for different gas turbine components and using the characteristic maps for compressor and turbine. This physical model and the required control system are simulated in Matlab/Simulink environment. The performance of the developed model for dry engine is validated by real data from an operating plant. After dry engine validation, the steam injection is also considered in the physical model and finally, the proposed control algorithm is added to the developed model. The performance of the proposed control system is compared with the regular gas turbine control algorithm in two different scenarios.

In first simulation, the performance of both control systems is compared to a step change in gas turbine load when the gas turbine is running in the island mode. The simulation results and important engine variables are shown in Figure 6. The results are summarized in the following paragraphs.

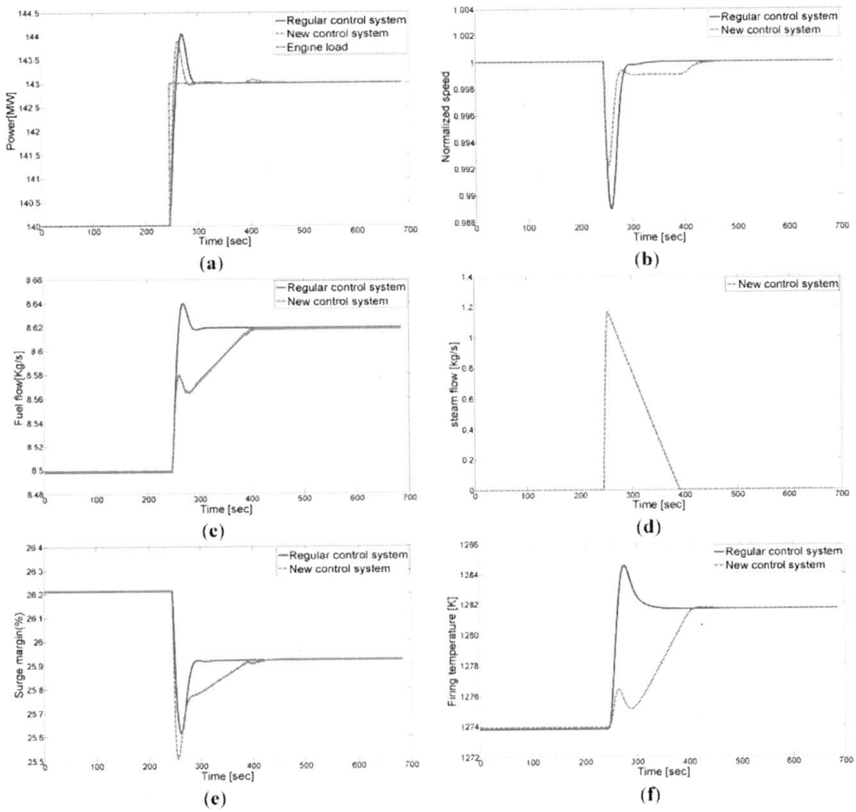

Figure 6. Comparison between the new control system and regular gas turbine control system in island mode-steam injection controller setting (dead band upper limit and lower limit = 0.001, −∞, gain = 20, saturation limit = 0.3): (**a**) power; (**b**) speed; (**c**) fuel flow; (**d**) steam flow; (**e**) surge margin; and (**f**) firing temperature.

The response of the new control system to the step change load is faster than the original control algorithm. This results in a smaller frequency drop for the new controller and the peak power of the new controller is also smaller than the original control algorithm. On the other hand, the new control system has a small speed deviation for a short period of time after frequency recovery. This deviation is related to the hold function in new controller structure. The hold function reduces the amount of steam injection slowly after the frequency recovery and the governor tries to keeps the output power constant by increasing the fuel flow. Since the governor works with the speed deviation, there should be a small error, therefore, the governor can replace the decrease in steam injection with the increase in fuel flow.

As stated in previous sections, the steam injection can reduce the compressor surge margin and it affects the surge margin more than the engine overfueling. Moreover, the surge margin is also decreased by a speed reduction. Consequently, although steam injection reduces the surge margin, it will result in a smaller frequency dip. Therefore, the compressor surge margin is only slightly affected by the new control algorithm.

A small amount of steam injection can enhance the gas turbine performance. The maximum steam mass flow rate is 1.2 kg/s and the total amount of injected steam during this transient maneuver is about 86.5 kg. Also, application of the steam during the transient maneuvers reduces the temperature gradient in firing temperature and consequently in engine components, which will increase the service life of the engine components.

The second simulation investigates the maximum allowable change load for these control algorithms in island mode. There are two main restrictions that limit the allowable change load that the gas turbine can have. Firstly, the temperature controller, which reduces the fuel flow when the firing temperature reaches to its safety limits, makes it impossible to load the gas turbine further. This limitation usually plays a role in high engine loads where the engine already has a high firing temperature, and thus it will be activated with a small amount of overfueling. The second limitation is the safety limit for surge margin which plays the main role at lower loads. In these loads the firing temperature is relatively low and this means that there is a large gap before the temperature controller is activated. Consequently, the engine can be overfueled more and this large overfueling will push the engine towards the surge region before the temperature controller is activated. Considering both of these limitations the performance of the proposed control algorithm (for different amount of maximum allowable steam injection) *vs.* the original control algorithm is presented in Figure 7.

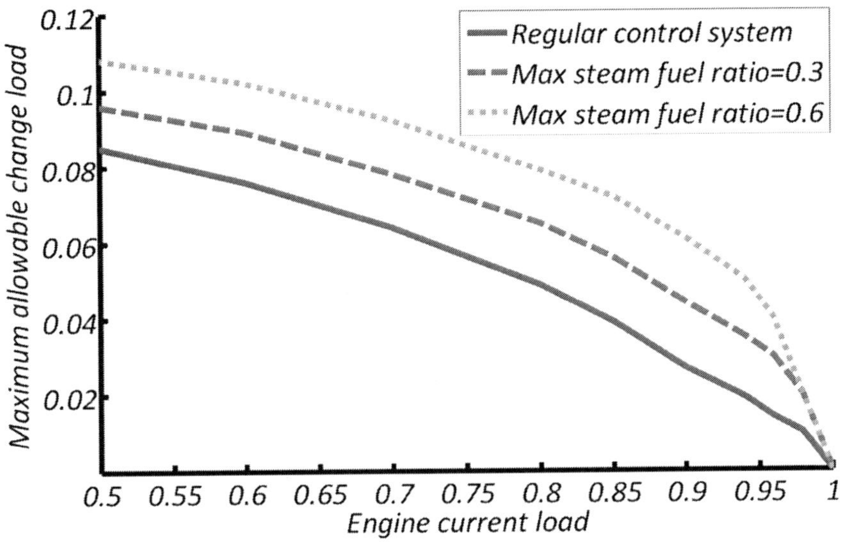

Figure 7. Maximum allowable change load in different load for the proposed control algorithm *vs.* the regular control algorithm when the gas turbine is running in island mode.

It can be seen that engine allowable change load will be decreased when the engine gets closer to its full load condition. Additionally, in high engine loads, where the temperature controller limits the maximum allowable change load, the steam injection can enhance the gas turbine performance. This is because steam injection will reduce the firing temperature and therefore, the temperature controller will not be activated very quickly. Also, steam injection augments the output power without increasing the firing temperature. On the other hand, at low engine loads, the surge margin limits the allowable change load. Although steam injection reduces the compressor surge margin, but when the control system takes advantage of steam injection, the engine can respond much faster to sudden change loads. Consequently, the engine will experience a smaller frequency drop and the compressor surge margin decrease will be less than the regular control algorithm because of speed reduction. This effect not only compensates the surge margin reduction because of steam injection, but also overrides that effect in large change loads which will lead to large frequency drops. Therefore, the engine will be able to manage higher change loads.

Another important issue in gas turbine performance is the engine maximum loading rate near full load condition. Faster engine loading means more overfueling and since the firing temperature is high near full load condition, more overfueling will activate the temperature controller and limits the gas turbine loading rate. Therefore, most of the gas turbine manufactures decrease the engine loading rate near full load condition. The proposed control algorithm can enhance the maximum engine loading rate near full load condition as well. For example, for the sample engine, the maximum loading rate is 4.6 MW/min for regular control algorithm and it can be increased to 8 MW/min and 10.5 MW/min with the proposed control algorithm for maximum steam fuel ratio of 0.3 and 0.6, respectively.

These simulation results show that the proposed control algorithm can improve the performance of the gas turbine during the frequency dips and transient maneuvers. The proposed control algorithm will be more effective near full load condition where the regular control algorithm is not capable to manage even small frequency dips.

6. CONCLUSIONS

In this paper, a new control system is developed to overcome the shortcomings of the gas turbine regular control algorithm in handling large frequency dips or frequency dips near full load condition. The proposed control algorithm takes advantage of steam injection into the combustion chamber during frequency dips to help the gas turbine regular control algorithm. The required control loop is developed and some simulations have been done to investigate the performance of the proposed control algorithm. The main results can be summarized as follows:

- A small amount of steam injection can enhance the gas turbine performance if the machine is subjected to a step change in engine load. The application of steam injection will lead to faster response and as a result smaller frequency drop. It would also reduce the temperature gradient and peak temperature in the combustion chamber.
- The new control algorithm can improve the maximum allowable engine change load, especially near full load condition. This is due to the fact that steam injection reduces the firing temperature in the combustion chamber, so it can postpone the activation of the temperature controller which will prevent the turbine from being loaded.

- Steam injection during frequency dips can improve the engine loading rate near full load condition significantly by reducing the firing temperature.
- Application of the new control algorithm affects the surge margin only slightly. This is because what we lose in surge margin because of steam injection will be gained due to the smaller frequency dip.

The current work only investigates the possibility of event base steam injection to enhance the performance of the gas turbine. The optimum amount of steam injection for best performance is the issue that can be investigated in future works.

Appendix A: Governor and Temperature Controller Details

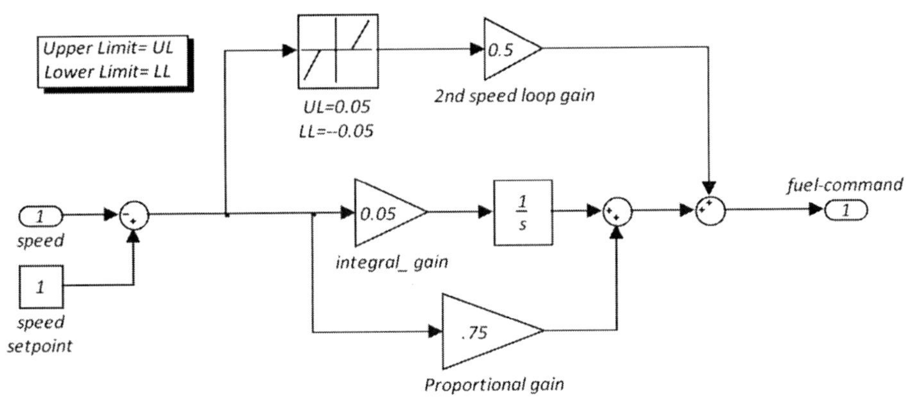

Figure A1. Governor (island mode).

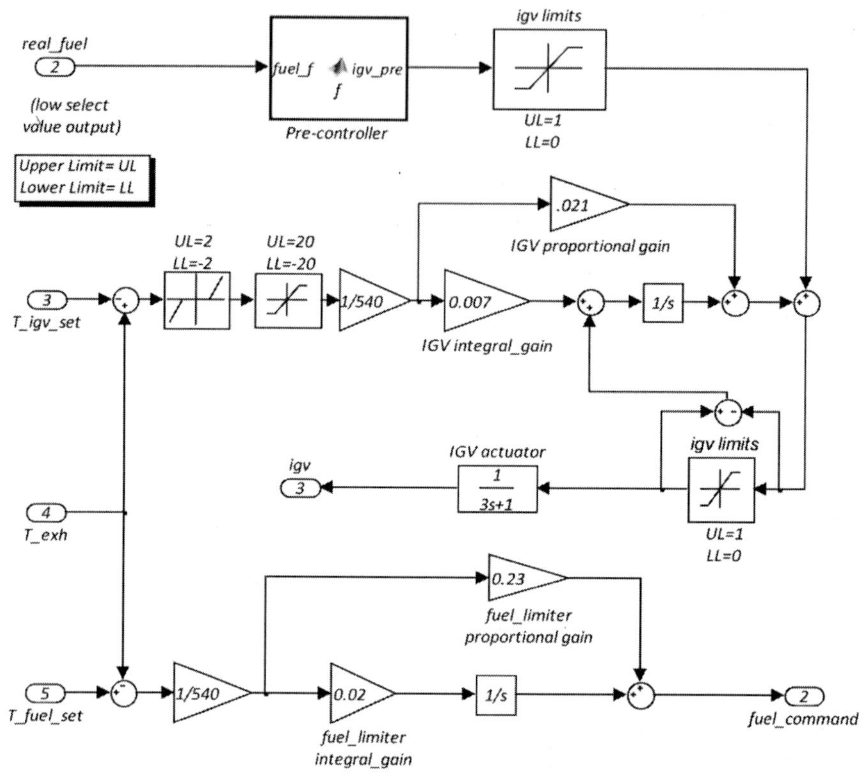

Figure A2. Temperature controller.

REFERENCES

1. Mello, F.P.; Ahner, D.J. Dynamic models for combined cycle plants in power system studies. *IEEE Trans. Power Syst.* **1994**, *9*, 1698–1708.
2. Suzaki, S.; Kawata, K.; Sekoguchi, M.; Goto, M. Mathematical Model for a Combined Cycle Plant and Its Implementation in an Analogue Power System Simulator. In Proceedings of the IEEE Power Engineering Society Winter Meeting, Singapore, 23–27 January 2000; Volume 1, pp. 416–421.
3. Camacho, E.F.; Samad, T.; Garcia-Sanz, M.; Hiskens, I. Control for Renewable Energy and Smart Grids. In *The Impact of Control Technology*; Samad, T., Annaswamy, A.M., Eds.; IEEE Control Systems Society: New York, NY, USA, 2011; pp. 69–88.

4. Lalor, G.; Ritchie, J.; Flynn, D.; O'Malley, M.J. The impact of combined-cycle gas turbine short-term dynamics on frequency control. *IEEE Trans. Power Syst.* **2005**, *20*, 1456–1464.
5. Walsh, P.P.; Fletcher, P. *Gas Turbine Performance*, 2nd ed.; Blackwell Science Ltd.: Oxford, UK, 2006.
6. Kakimoto, N.; Baba, K. Performance of gas turbine-based plants during frequency drops. *IEEE Trans. Power Syst.* **2003**, *18*, 1110–1115.
7. Jonsson, M.; Yan, J. Humidified gas turbines—A review of proposed and implemented cycles. *Energy* **2005**, *30*, 1013–1078.
8. Lindquist, T.; Rosén, P.M.; Torisson, T. *Theoretical and Experimental Evaluation of the EvGT-Process*; Lund Institute of Technology: Lund, Sweden, 2000; pp. 457–469.
9. Rice, I.G. Steam-Injected Gas Turbine Analysis—Part I—Steam Rates. In Proceedings of the International Gas Turbine and Aeroengine Congress and Exposition, Cincinnati, OH, USA, 24–27 May 1993.
10. Rice, I.G. Steam-Injected Gas Turbine Analysis—Part II—Steam Cycle Efficiency. In Proceedings of the International Gas Turbine and Aeroengine Congress and Exposition, Cincinnati, OH, USA, 24–27 May 1993.
11. Rice, I.G. Steam-Injected Gas Turbine Analysis—Part III—Steam Regenerated Heat. In Proceedings of the International Gas Turbine and Aeroengine Congress and Exposition, Cincinnati, OH, USA, 24–27 May 1993.
12. Lalor, G.; O'Malley, M.J. Frequency Control on an Island Power System with Increasing Proportions of Combined Cycle Gas Turbines. In Proceedings of the IEEE Power Tech Conference, Bologna, Italy, 23–26 June 2003.
13. Razak, A.M.Y. *Industrial Gas Turbines Performance and Operability*; CRC Press: Cambridge, UK, 2007.
14. Jurado, F.; Carpio, J. Improving distribution system stability by predictive control of gas turbines. *Energy Convers. Manag.* **2006**, *47*, 2961–2973.
15. Correa, S.M. A review of NO_x Formation under gas-turbine combustion conditions. *Combust. Sci. Technol.* **1993**, *87*, 329–362.

16. Dryer, F.L. Water addition to practical combustion systems—Concepts and applications. *Symp. Int. Combust.* **1976**, *16*, 279–295.

17. Schorr, M.M. NO_x Emission Control for Gas Turbines: A 1992 Update on Regulations and Technology. In Proceedings of the ASME Cogen-Turbo, Houston, TX, USA, 1–3 September 1992.

18. Roumeliotis, I.; Mathioudakis, K. Evaluation of water injection effect on compressor and engine performance and operability. *Appl. Energy* **2010**, *87*, 1207–1216.

19. Haselbacher, H. Performance of water/steam injected gas turbine power plants consisting of standard gas turbines and turbo expanders. *Int. J. Energy Technol. Policy* **2005**, *3*, 12–23.

20. Saravanamuttoo, H.I.H.; Rogers, G.F.C.; Cohen, H. *Gas Turbine Theory*, 5th ed.; Pearson Education: London, UK, 2001.

3
CHAPTER

PARAMETRIC ANALYSIS OF A TWO-SHAFT AERODERIVATE GAS TURBINE OF 11.86 MW

R. Lugo-Leyte [1], M. Salazar-Pereyra [2,*], H. D. Lugo Méndez [1], I. Aguilar-Adaya [1], J. J. Ambriz-García [1] and J. G. Vázquez Vargas [3]

[1] *Universidad Autónoma Metropolitana-Iztapalapa, Departamento de Ingeniería de Procesos e Hidráulica, Av. San Rafael Atlixco No. 186, Col. Vicentina, 09340, Iztapalapa, Ciudad de México, D.F., Mexico*

[2] *Tecnológico de Estudios Superiores de Ecatepec, División de Ingeniería Mecatrónica e Industrial, Av. Tecnológico Esq. Av. Hank González, Col. Valle de Anáhuac, 55210, Ecatepec, Estado de México, Mexico*

[3] *MAVAZCAR S.A. de C.V. Suministros, Servicios e Ingeniería*

ABSTRACT

The aeroderivate gas turbines are widely used for power generation in the oil and gas industry. In offshore marine platforms, the aeroderivative gas turbines provide the energy required to drive mechanically compressors, pumps and electric generators. Therefore, the study of the performance of aeroderivate gas turbines based on a parametric analysis is relevant to carry out a diagnostic of the engine, which can lead to operational as well as predictive and/or corrective maintenance actions. This work presents a methodology based on the exergetic analysis to estimate the irrevesibilities and exergetic efficiencies of the main components of a two-shaft aeroderivate gas turbine. The studied engine is the Solar Turbine Mars 100, which is rated to provide 11.86 MW. In this engine, the air is compressed in an axial compressor achieving a pressure ratio of 17.7 relative to ambient conditions and a high pressure turbine inlet temperature of 1220 °C. Even if the thermal efficiency associated to the pressure ratio of 17.7 is 1% lower than the maximum thermal efficiency, the irreversibilities related to this pressure ratio decrease approximately 1 GW with respect to irreversibilities of the optimal pressure ratio for the thermal efficiency. In addition, this paper contributes to develop a mathematical model to estimate the high turbine inlet temperature as well as the pressure ratio of the low and high pressure turbines.

Keywords: aeroderivate gas turbine; exergetic efficiency; exergy; irreversibility

1. INTRODUCTION

Gas turbines are used successfully to power aircraft as well as to generate power in industrial applications. In offshore marine platforms, gas turbines are used as stationary power plants or as prime mover of pumps and compressors, because they are compacts with low weigh and reliable operation. There exist aero-derivate gas turbines of low and medium power, and their maximum capacity of generation is of 65 MW, depending on their compressor pressure ratio, high pressure turbine inlet temperature and of their shaft number [1,2]. In addition, the performance of the delivered power of a gas turbine is affected by the conditions presented at the place where it is installed, mainly ambient temperature, atmospheric pressure and the air relative humidity [3].

A two-shaft aero-derivate gas turbine is composed by a gas generator and a free power turbine also known as a low pressure turbine. The gas generator consists of a compressor, a combustion chamber and a high pressure turbine. The gas generator turbine provides power for the air compressor and the free power turbine generates the useful power output. Likewise, improvements in the design and operation of gas turbines have come along with advances in aerodynamics, thermodynamics and metallurgy has made possible for today's gas turbines to withstand temperatures in the 1415 °C range, compressor pressure ratios up to 34:1 and overall thermal efficiency up to 42% [4,5].

However, as the gas turbines present large losses energetics and economics in spite of the technologic development, Bejan [6] has development a useful tool that supply information about performance of conversion systems of energy, and it gives the possible upgrade of the devices, supporting with a parametric analyses. The first law of thermodynamic is the principal tool to do thermodynamic analyses, but it only give responses about of thermal efficiency of the cycle and doesn't provide the irreversibilities of each of components of the system. Along with the exergetic analyses, the quality of the energy and the irreversibilities of each of the components of the system can also be assessed. The exergetic analysis use the mass and energy balances and the second law of thermodynamics to do the analysis, design and the improvement of the energetic systems; moreover, it identifies the types and magnitude of the wastes, losses, and efficiencies of system to establish the potential of improvement of energy systems [6,7,8,9].

The literature presents several works about thermodynamics analyses of gas turbines, but most are focused to study the performance of simple gas turbines or power plants of cycle combined [10,11,12,13,14,15,16,17]. Nevertheless, most works only study one condition of operation, with the exception of [3], which establishes that the turbine inlet temperature, the drops of pressure, environmental temperature and compressor pressure ratio are parameters that change the thermal efficiency of gas turbines.

Hakan Aydin [1,2] presents an exergetic study of an aero-derivate gas turbine LM6000, Hakan reports an exergetic efficiency of 39% for a compressor pressure ratio of 30 and a power of 43 MW. He has reported that the value of irreversibilities in the combustion chamber is 25.91 MW; it represents 60% of the generated power and its exergetic efficiency is 85.1%, considering a drop of pressure of 5%. However, it is not possible to do a comparison that only presents the analysis for a condition of operation. He also presents that the destruction of exergy of the turbine of gas generator is bigger than free power turbine. Nevertheless, he doesn't establish that the aero-derivate gas turbine has a system of Nozzle Guide Vane (NGV) and that it increases the high pressure turbine inlet temperature.

The aim of this work is to present a methodology based on the exergetic analysis to estimate irrevesibilities and exergetic efficiencies of each of the main components of an aero-derivate gas turbine. The studied engine is the Mars 100, an aeroderivative gas turbine with two-shaft design and rated to provide 11.86 MW. In this engine, the air is compressed in an axial compressor achieving a pressure ratio of 17.7 relative to ambient conditions and a high pressure turbine inlet temperature of 1220 °C. The compressor and turbine efficiencies are assumed to be 0.85 and 0.80, respectively.

2. METHODOLOGY

A two-shaft aeroderivate gas turbine is conformed by a gas generator and a free power turbine also known as low pressure turbine (LPT), as depicted in the schematic diagram of Figure 1a. The gas generator consists of a compressor, a combustion chamber and a high pressure turbine (HPT). An aeroderivative gas turbine operates on the thermodynamic cycle presented in the temperature-entropy diagram of Figure 1b, in which air entering the compressor at state 1 is compressed to some higher pressure and temperature at state 2. Leaving the compressor, air enters the combustion chamber, where combustion occurs by fuel injection. In the combustion process, a pressure drop occurs by the mixing, burning and cooling phenomena. The exhaust gases leave the combustion system and enter the HPT at state 3. In the HPT of the gas generator, the hot gases are expanded to state 4 and their thermal energy is converted into work, which is used to drive the compressor of the gas generator. The expanded gases at state 4 enters the LPT coupled aerodynamically to the gas generator, where they are expanded to state 5, at a pressure higher than the air pressure at state 1 to generate the useful power output.

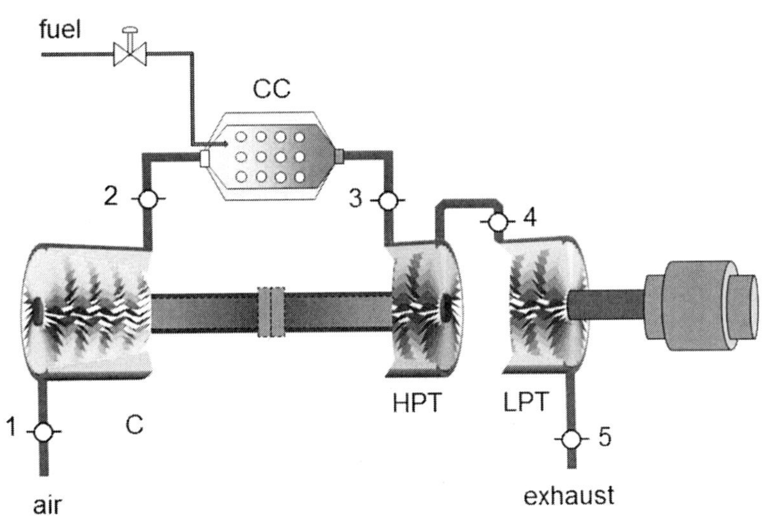

(a) Schematic diagram.

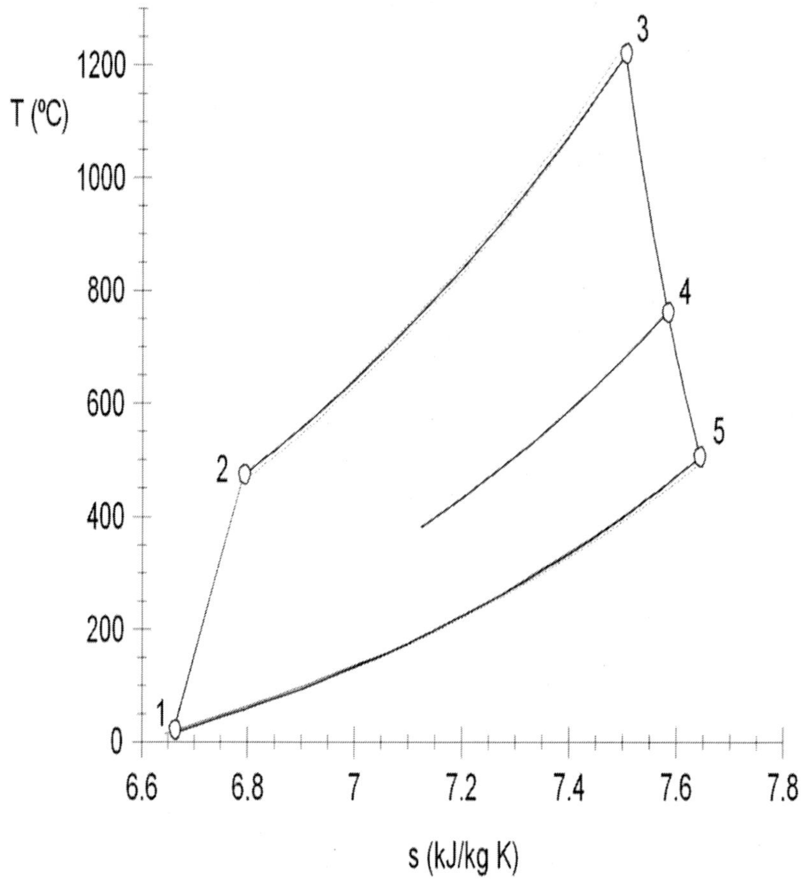

(b) Temperature entropy diagram.

Figure 1. Two-shaft gas turbine.

In this work, the development of a systematic methodology is pursued to evaluate the performance of the Mars 100 gas turbine, based on the estimation of the engine component irreversibilities. The Solar Mars 100 is an aeroderivative two-shaft gas turbine rated to provide a Standard conditions base load of 11.86 MW and designed especially for mechanical drive applications. Its main components are a 15-stage axial-flow air compressor with a compression ratio of 17.7:1, an annular combustion chamber, a two-stage reaction HPT and a two-stage axial flow LPT. Since the engine summary screen of the Mars 100 only reports pressures and temperatures at the states one, two and five, a parametric analysis of the gas turbine performance becomes relevant, taking the compressor pressure ratio and the HPT inlet temperature T4, as the main parameters, whose studied values are taken from the manufacturer operational recommendations. This analysis considers the existence of pressure drops in the combustion chamber as well as in the LPT and is carried out for compressor and turbine isoentropic efficiencies of 0.85 and 0.80, respectively.

The compressor pressure ratio is the relation between the compressor discharge pressure, P2, and the compressor suction pressure, P1,

$$\pi_C = \frac{P_2}{P_1} \tag{1}$$

For the actual compression process, the specific input compression work is

$$w_C = \frac{c_{Pa} T_1}{\eta_C} \left(\pi_C^x - 1 \right) \tag{2}$$

The compression work, given by Equation (1), is a function of the environmental parameter T1, and the technological parameters π C and η C .

The increase of entropy for a compression process can be obtained from the following expression

$$s_2 - s_1 = c_{Pa} \ln\left(1 + \frac{\pi_C^x - 1}{\eta_C} \right) - R_a \ln \pi_C \tag{3}$$

In the combustion process, a pressure drop of 5% in considered, and the amount of heat transferred to the air at constant pressure is a function of the temperatures ratio y=T 1 /T 3 , this relation involves an environmental and technological conditions. The technological temperature T 3 is constrained by the technological development of the material gas turbine, while T 1 is constrained by the variations in the environmental conditions.

$$q_{in} = c_{Pg}T_1\left[y - 1 - \frac{1}{\eta_C}\left(\pi_C^x - 1\right)\right]$$

(4)

The entropy change in the heating process is given by

$$s_3 - s_2 = c_{Pg}\ln\left(\frac{y}{1 + \frac{1}{\eta_C}\left(\pi_C^x - 1\right)}\right) - R_g\ln\left(1 - \frac{\Delta P_{cc}}{\pi_C P_1}\right)$$

(5)

The compressor work is provided by the work generated by the HPT, and the losses due to the transmission of work from turbine to compressor are determined by introducing the mechanical efficiency of η m =98% .

$$w_C = \eta_m w_{HPT}$$

(6)

The outlet temperature of the HPT can be expressed in terms of y,η SC ,η m ,π x C ,c pa c pg , as follows

$$T_4 = T_3 - \frac{c_{Pa}}{c_{Pg}}\frac{T_1}{\eta_C\eta_m}\left(\pi_C^x - 1\right)$$

(7)

The expression of the HPT pressure ratio in terms of the temperature ratio T 4 T 3 is given by

$$\left(\frac{1}{\pi_{HPT}^x}\right) = 1 - \frac{c_{Pa}}{c_{Pg}}\frac{1}{y}\frac{1}{\eta_{HPT}\eta_m\eta_C}\left(\pi_C^x - 1\right)$$

(8)

For an expansion process the change in entropy can be expressed as

$$s_4 - s_3 = c_{Pg}\ln\left(1 - \frac{c_{Pa}}{c_{Pg}}\frac{1}{\eta_{HPT}\,\eta_C\,\eta_m\,y}\left(\pi_C^x - 1\right)\right) - R_g\ln\left(\frac{1}{\pi_{HPT}}\right)$$

(9)

The unknown temperature and pressure at state four, T4 and P4, can be found from the LPT pressure ratio, πLPT; and the useful output specific work generated by the power turbine can then be written as follows

$$w_{LPT} = c_{P_g} \eta_{LPT} T_4 \left(1 - \frac{1}{\pi_{LPT}^x} \right) \quad (10)$$

The change in entropy for the expansion process in the LPT is given by

$$s_5 - s_4 = c_{Pg} \ln\left(1 - \eta_{LPT} \left(1 - \frac{1}{\pi_{LPT}^x} \right) \right) - R_g \ln\left(\frac{1}{\pi_{LPT}} \right) \quad (11)$$

The thermal efficiency is the relation between the actual work delivered by the free power turbine and the supplied heat by the fuel combustion,

$$\eta_{TH} = \frac{\eta_{LPT} T_4 \left(1 - \frac{1}{\pi_{LPT}^x} \right)}{T_1 \left[y - 1 - \frac{1}{\eta_C} \left(\pi_C^x - 1 \right) \right]} \quad (12)$$

The air flow rate required to generate a demanded power output can be then expressed as

$$\dot{m}_a = \frac{\dot{W}}{w_{tp}} \quad (13)$$

The fuel flow fuel is given by

$$\dot{m}_f = \frac{\dot{W}}{LHV \, \eta_{TH}} \quad (14)$$

By definition, the heat rate (HR) is the energy required to generate one kW-h, and its expression is given by:

$$HR = \frac{3600 \, \mathrm{T}_1 \left[y - 1 - \frac{1}{\eta_C} \left(\pi_C^x - 1 \right) \right]}{\eta_{LPT} T_4 \left(1 - \frac{1}{\pi_{LPT}^x} \right)} \quad (15)$$

3. EXERGETIC ANALYSIS

Exergy analysis is based on the second law of thermodynamics. This analysis is an essential tool to provide insights of the performance of energy conversion systems, and highlights the possible improvements. The exergy analysis has been extending to many applications, considering also some energetic, economic, and environmental factors [6,7,8]. The exergy is then a powerful tool for understanding and improving the sustainability of processes and systems.

The exergy of a heat transfer process for a given control surface is determined from the maximum work that could be obtained from it using the environment as a reservoir of zero grade thermal energy [7]. For a heat transfer rate Q . r and a temperature at the control surface where the heat transfer is taking place Tr, the maximum rate of conversion from thermal energy to work is

$$\dot{W}_{max} = \dot{E}^{Q_r} = \dot{Q}_r \tau = \dot{m}_f \, LHV \, \tau \tag{16}$$

where τ is the dimensionless exergetic temperature and is equal to the Carnot coefficient when the environment at T0 is used as the other thermal energy reservoir, and it can be expressed as

$$\tau = 1 - \frac{T_0}{AFT} \tag{17}$$

The Carnot coefficient in terms of the combustion temperature is presented in Figure 2, and this figure shows that the maximum work is approximately 88% considering the adiabatic flame temperature for natural gas as fuel. The losses due to the combustion process are approximately 12%; nevertheless the combustion gas temperature decreases for a high pressure turbine inlet temperature of the gas turbine Mars 100 of 1200 °C, leading to an additional loss of 8%, due to the cooling of the combustion gases, and these losses increase with the decreasing of T3.

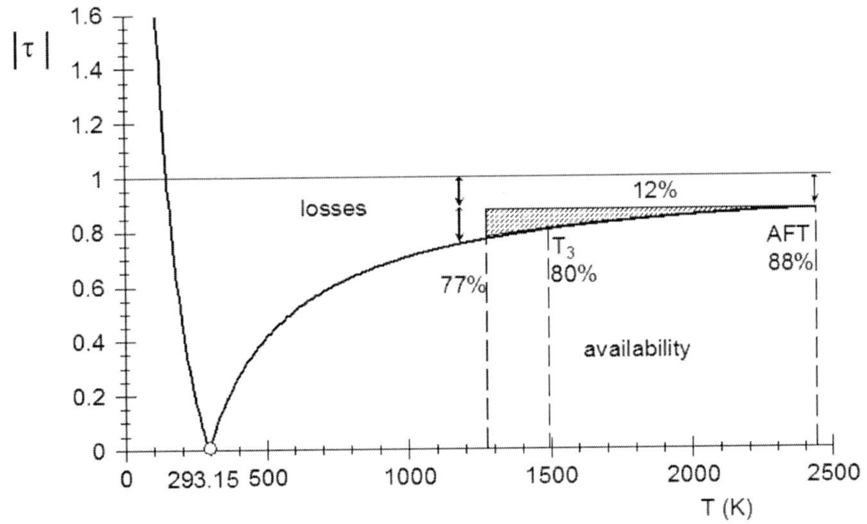

Figure 2. Exergetic temperature factor.

In order to obtain the specific physical exergy of the thermodynamic states of the aeroderivate gas turbine, the following expression is used:

$$\varepsilon = (h - h_0) - T_0 (s - s_0) \tag{18}$$

The specific physical exergy for air and combustion gases with constant specific heat is obtained from:

$$\varepsilon = c_P (T - T_0) - T_o \left(c_P \ln \frac{T}{T_0} - R \ln \frac{P}{P_0} \right) \tag{19}$$

Figure 3a,b present the control volumes of the expansion and compression processes. These figures show the exergies, works and irreversibilities of the inlets and outlets of the control volumes. The exergetic balances for the control volumes of compression and expansion processes are given by:

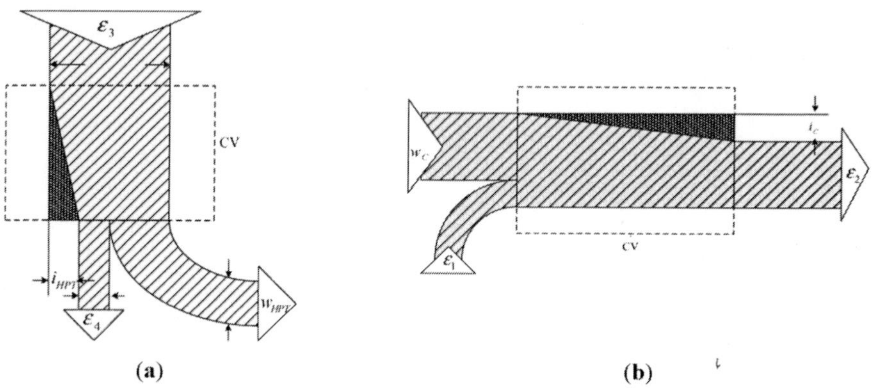

Figure 3. (a) Expansion process and (b) compression process.

3.1. Gas Turbine

$$\varepsilon_3 - \varepsilon_4 = w_{HPT} + i_{HPT} \tag{20}$$

The exergetic efficiency of the high pressure turbine in terms of the irreversibility and turbine efficiency is given by

$$\varphi_{HPT} = \frac{\eta_{HPT}}{\frac{i_{HPT}}{r_{HPT}} + \eta_{HPT}\left(1 - \frac{i_{HPT}}{r_{HPT}}\right)} \tag{21}$$

where rHPT, is the frictional reheat, which is defined as r HPT =h 4 −h 4s .

3.2. Compressor

The exergy balance for the compressor can be written as:

$$\varepsilon_1 + w_C = \varepsilon_2 + i_C \tag{22}$$

The exergetic efficiency of the compressor in terms of the irreversibility and compressor efficiency results as follows

$$\varphi_C = 1 - \frac{i_C}{r_C}(1 - \eta_C) \tag{23}$$

where, rC is the frictional reheat of the compression process, which is defines as r C =h 2 −h 2s .

Figure 4a presents the variations of the exergetic efficiencies of the compressor and high pressure turbine as a function of the ratio of frictional reheat and irreversibility for different values of the compressor efficiencies. The results show that the rC/iC increases as the compressor efficiency decreases due the increase of the irreversibilities in the compressor and the frictional reheat; likewise, by decreasing the compressor efficiencies and also decreasing rHPT/iHPT, since the increase of the irreversibilities is higher than the increase the frictional reheat. For a compressor efficiency of 0.86, the value of rC/iC is of 2.27 and the value exergetic efficiency of the compressor of 0.938. Indeed, a 0.84% reduction in the compressor efficiency can be observed that the rC/iC is increased 0.98%, due that the frictional reheat increase 17.01% and the irreversibilities only 15.8%; nevertheless, the exergetic efficiency of the high pressure turbine is not affected in great measure. As result of an increase of 7% of the irreversibilities in the high pressure turbine, the rHPT/iHPT decreases 1.1%.

Figure 4b presents the variations of the exergetic efficiencies of low and high pressure turbines as function of r/i for different values of the turbine efficiency. The results show that the rLPT/iLPT increases as the compressor efficiency decreases due to the increase of the irreversibilities in the low pressure turbine and the frictional reheat; likewise, by decreasing the turbine efficiencies and also decreasing rHPT/iHPT, since the increase of the irreversibilities (iHPT) is higher than the increase the frictional reheat (rHPT). For a turbine efficiency of 88%, the value of rLPT/iLPT is of 2.5795 and the value exergetic efficiency of the low pressure turbine of 0.9498. Indeed, reducing the turbine efficiency from 88% to 86% the rLPT/iLPT increases 1.07%, due that the frictional reheat increase 13.98% and the irreversibilities 12.77%. Nevertheless, the exergetic efficiency of the high pressure turbine decreases 0.73% and the rHPT/iHPT decreases 1.209%, a result of the increase of 20.84% in the irreversibility in the high pressure turbine.

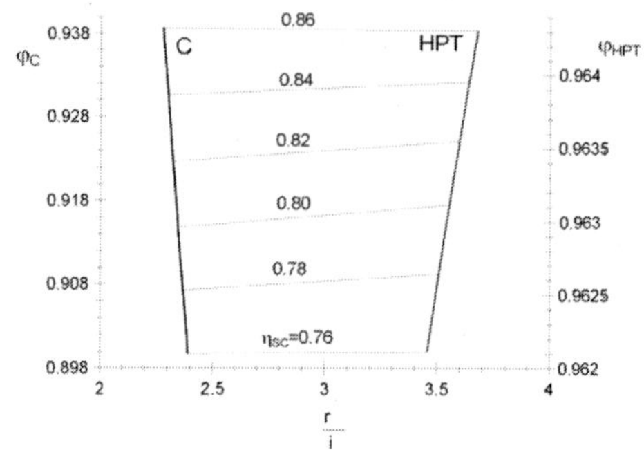

(**a**) Variation of exergetic efficiency of compressor and exergetic efficiency of high pressure turbine as function of ratio of frictional reheat and irreversibility for different compressor efficiencies.

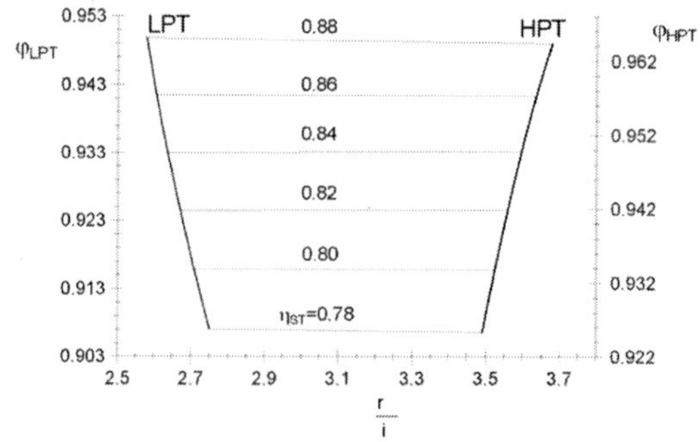

(**b**) Variation of exergetic efficiency of low pressure turbine and exergetic efficiency of high pressure turbine as function of ratio of reheat degree and irreversibility for different turbine efficiencies.

Figure 4. Variation of exergetic efficiencies of compressor and low and high pressure turbines as function of ratio of reheat degree and irreversibility for different compressor and turbine efficiencies.

In this analysis, it is supposed that the combustion process is adiabatic, and the irreversibilities due to the friction and the mixed are negligible.

The exergy balance for the combustion chamber is the following:

$$\varepsilon_2 + \varepsilon_{AFT} = \varepsilon_3 + i_{cc} \tag{24}$$

The exergy delivered by the fuel to the adiabatic flame temperature, (ε AFT), it can be expressed as

$$\varepsilon_{AFT} = \Delta g_r + R_f T_a \ln \frac{P_f}{P_a} \tag{25}$$

The exergetic efficiency of the combustion is given by

$$\varphi_{cc} = \frac{\varepsilon_3}{\varepsilon_2 - \varepsilon_{AFT}} \tag{26}$$

where ε AFT is negative, due to the amount of the Gibbs free energy.

It will perform a parameter study to determine how the system performance varies with different operating parameters. Table 1 shows the mathematic models to assess the work, supplied heat, irreversibilities, and exergetic efficiencies of the engine components.

Table 1. Work, supplied heat, irreversilities and exergetic efficiencies of the engine components.

Engine Components	Energy	Irreversibility	Exergetic Efficiency
Compressor	$w_C = \frac{c_{pa}T_1}{\eta_C}(\pi_C^x - 1)$	$i_C = \varepsilon_1 - \varepsilon_2 + w_C$	$\varphi_C = \frac{\varepsilon_2 - \varepsilon_1}{w_C}$
Combustion chamber	$q_{in} = c_{Pg}T_1\left[y - 1 - \frac{1}{\eta_C}(\pi_C^x - 1)\right]$	$i_{cc} = \varepsilon_2 - \varepsilon_3 + \varepsilon_f$	$\varphi_{cc} = \frac{\varepsilon_3}{\varepsilon_2 - \varepsilon_f}$
High pressure turbine	$w_{HPT}\eta_m = w_C$	$i_{HPT} = \varepsilon_3 - \varepsilon_4 - w_{HPT}$	$\varphi_{HPT} = \frac{w_{HPT}}{\varepsilon_3 - \varepsilon_4}$
Low pressure turbine	$w_{LPT} = c_{Pg}\eta_{LPT}T_4\left(1 - \frac{1}{\pi_{LPT}^x}\right)$	$i_{LPT} = \varepsilon_4 - \varepsilon_5 - w_{LPT}$	$\varphi_{LPT} = \frac{w_{LPT}}{\varepsilon_4 - \varepsilon_5}$

Figure 5 shows the diagram exergy-enthalpy. It presents the thermodynamic states of aeroderivate gas turbine. State three has the maximum exergy because it has the higher temperature and pressure of the system. The difference of enthalpies of the high pressure turbine is higher than the difference of enthalpies of the compressor by 2% due to the mechanical losses in the shaft. However, the difference of exergies of the high pressure turbine is 483.77 kJ/kg and the difference of exergies of the compressor of 421.32 kJ/kg. Then, the difference is 13%, 2% are for the mechanical losses in the shaft and 11% by internal irreversibilities in the high pressure turbine.

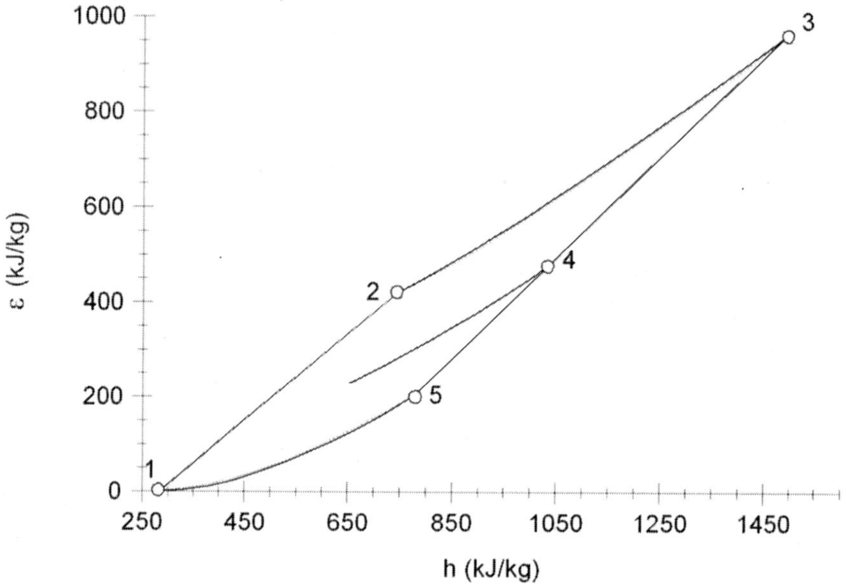

Figure 5. Exergy enthalpy diagram of a two-shaft aero-derivate gas turbine.

4. RESULTS AND DISCUSSION

Figure 6a shows the thermal efficiency as a function of work output specific for different values of high pressure turbine inlet temperature and different values of pressure ratios of the compressor. For a high pressure turbine inlet temperature of 1220 °C, the pressure ratio of the compressor to obtain the maximum thermal efficiency is 21 (point b). Nevertheless, the pressure ratio of the compressor of operation of turbine Mars 100 is 17.7 (point a), hence there is a depletion of 1% of the thermal efficiency and 15 kJ/kg of the specific work output. It shows that by decreasing the high pressure turbine inlet temperature 100 °C keeping constant pressure ratio, the specific work output decreases 48.33 kJ/kg and the thermal efficiency rises 2.24%.

Figure 6b shows the fuel flow rate and the thermal efficiency as a function of pressure ratio of compressor for different values of high pressure turbine inlet temperature. It must be noted that to generate 11.8 MW, the value pressure ratio that minimizes fuel flow rate is 21 (point b); nevertheless, the value pressure ratio of operation of turbine Mars 100 is 17.7 (point a), hence there is an increase of fuel flow rate of 1% (0.0078 kgf/s) with respect to point b. It shows that by decreasing the high pressure turbine inlet temperature to 1100 °C keeping constant pressure ratio, the fuel flow rate increases to the 0.577 kgf/s, supplying 10% more fuel to maintain the same power on Mars 100.

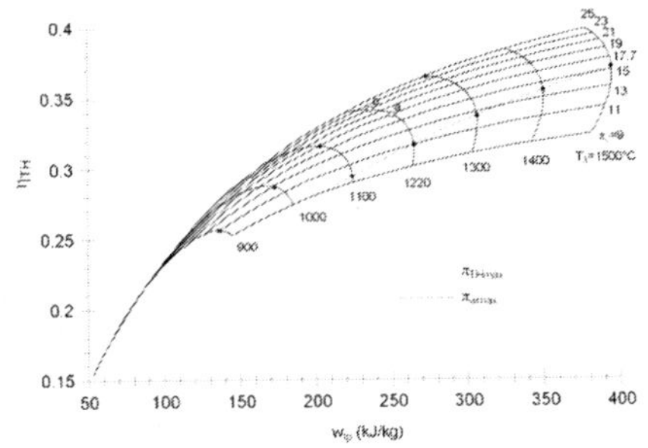

(**a**) Dependence of the thermal efficiency on the output specific for different high pressure turbine inlet temperatures and compression ratios of compressor.

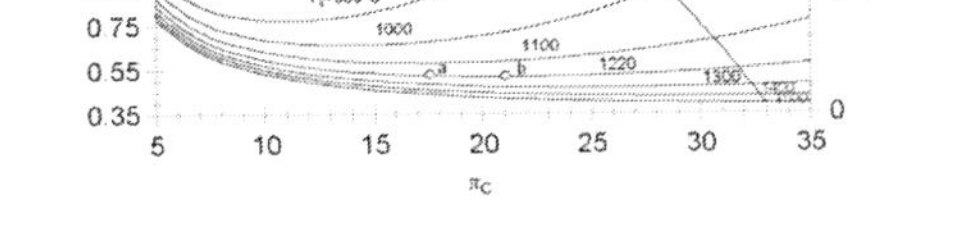

(**b**) Dependence fuel flow rate and thermal efficiency on compression ratios of compressor for different high pressure turbine inlet temperatures.

Figure 6. Design performance charts for the two-shaft aero-derivate gas turbine.

Figure 7 shows the effects the pressure ratio of the compressor on the heat rate and thermal efficiency for different values of the T3. The behavior of the Figure 7 is similar to the Figure 6b due to the relationship that there is with the fuel. For the operation conditions of the turbine Mars 100, the heat rate is 12,200 kJ/kW-h (point a); with respect to πC for maximum thermal efficiency (point b), there is a variation lower to 100 kJ/kW-h.

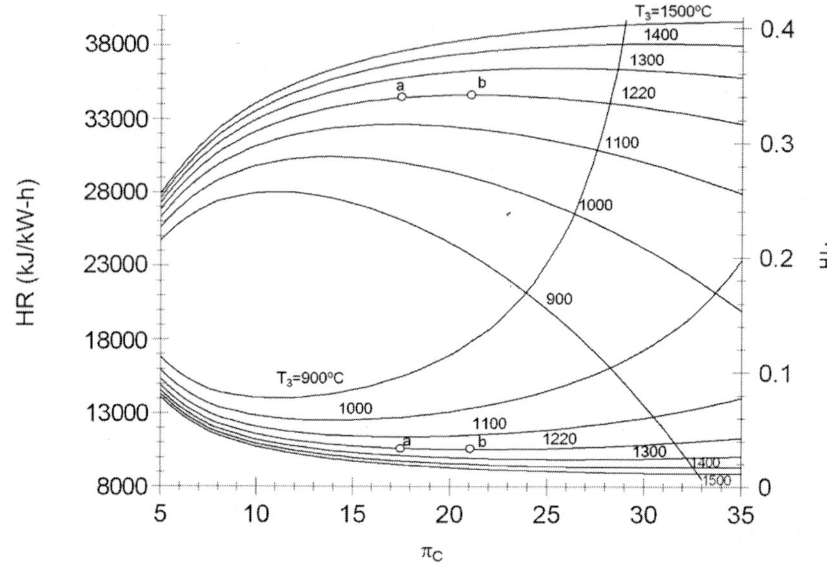

Figure 7. Heat rate and thermal efficiency against compression ratio of compressor for different high pressure turbine inlet temperatures.

It is important to analyze the thermodynamic performance of the compressor in relation to the compressor efficiency, a drop of the ηSC effects to all components of derivate gas turbine, the results of the exergetic analysis are presented in Table 2. From this table, an increment of $\dot{I}$ C , $\dot{I}$ CC , $\dot{I}$ HPT and $\dot{I}$ LPT of 20.26%, 1,5%, 7.5% and 0.14%, respectively, and the exergetic efficiencies of the low and high pressure turbines decreases in 1%, as well as ηC decreases from 86% to 84% for a turbine efficiency of 85% can be observed. The heat transferred to the aeroderivate turbine increases 26.01% when the ηT decreases from 86% to 76% to generate an output of 11.86 MW.

It can be assumed that the turbine has been being operating for ten years and that the compressor efficiency has dropped from 86% to 76%. As a result, its exergetic efficiency decreases from 94% to 90% and the irreversilities of the high pressure turbine are increased 51.77%.

Table 2. Variation of exergy of heat transfer, irreversibility rate and exergetic efficiencies of the aeroderivate gas turbine components as functions of the compressor efficiency for ηHPT = ηLPT = 0.85.

η_C	$\dot{E}^Q$ (kW)	$\dot{I}_C$ (kW)	$\dot{I}_{cc}$ (kW)	$\dot{I}_{HPT}$ (kW)	$\dot{I}_{LPT}$ (kW)	φ_C	φ_{HPT}	φ_{LPT}
0.86	20.190.14	1.112.28	7.919.59	912.20	816.08	0.94	0.95	0.94
0.84	20,963.22	1,337.69	8,038.82	980.60	814.95	0.93	0.95	0.94
0.82	21,845.64	1,590.75	8,178.14	1,059.22	813.73	0.92	0.95	0.94
0.80	22,862.27	1,877.49	8,342.24	1,150.43	812.39	0.91	0.95	0.94
0.78	24,046.11	2,205.87	8,537.40	1,257.41	810.91	0.91	0.95	0.94
0.76	25,441.93	2,586.66	8,772.14	1,384.46	809.29	0.90	0.95	0.94

The change of the irreversibility rate as a function of the turbine efficiency is shown in Table 3. From this table, an increment of $\dot{I}_C$, $\dot{I}_{CC}$, $\dot{I}_{HPT}$ and $\dot{I}_{LPT}$ of 4.73%, 4.07%, 26.6% and 18.15%, respectively, and the exergetic efficiencies of the low and high pressure turbines decreases in 1%, as well as ηT decreases from 88% to 86% for a compressor efficiency of 80% can be observed. The heat transferred to the aeroderivate turbine is increased by 26.01% when the ηT decreases from 88% to 86% to generate an output of 11.86 MW.

It is assumed that the turbine has been operating for ten years and that the turbine efficiency has dropped from 88% until 85%. As a result, the exergetic efficiency decreases 1% and the irreversibility of the high pressure turbine increases 10%.

Table 3. Change of the exergy rate of the heat transfer, irreversibilities rate and exergetic efficiencies of components of the aeroderivate gas turbine as functions of the turbine efficiency for ηC = 0.80.

η_T	$\dot{E}^Q$ (kW)	$\dot{I}_C$ (kW)	$\dot{I}_{cc}$ (kW)	$\dot{I}_{HPT}$ (kW)	$\dot{I}_{LPT}$ (kW)	φ_C	φ_{HPT}	φ_{LPT}
0.88	17,783.42	1.002.16	7.314.01	611.30	632.66	0.94	0.96	0.95
0.86	19,186.12	1.049.58	7.624.87	773.91	747.49	0.94	0.96	0.94
0.84	20,747.46	1.102.37	7,970.88	961.71	864.47	0.94	0.95	0.93
0.82	22,497.97	1.161.54	8.358.82	1,179.99	983.70	0.94	0.94	0.92
0.80	24,476.83	1.228.44	8.797.37	1,435.55	1.105.28	0.94	0.93	0.92

Figure 8a shows the exergetic efficiency of compressor as function of irreversibility rate in compressor for different high pressure turbine inlet temperatures and different pressure ratios of the compressor. This figure shows that by increasing the temperature inlet high pressure turbine, the irreversibility rates decrease. It presents that keeping constant the high pressure turbine inlet temperature and increasing the pressure ratio of compressor, the exergetic efficiency of the compressor and the irreversibilities rate in compressor are increased. Note, for instance, for the condition of point a, the exergetic efficiency is 91.7% and the irreversibility rate is 1842 kW; increasing the pressure ratio of compressor from 17.7 to 21 (point b), the exergetic efficiency and irreversibility rates are increased 1% and 4.58%, respectively.

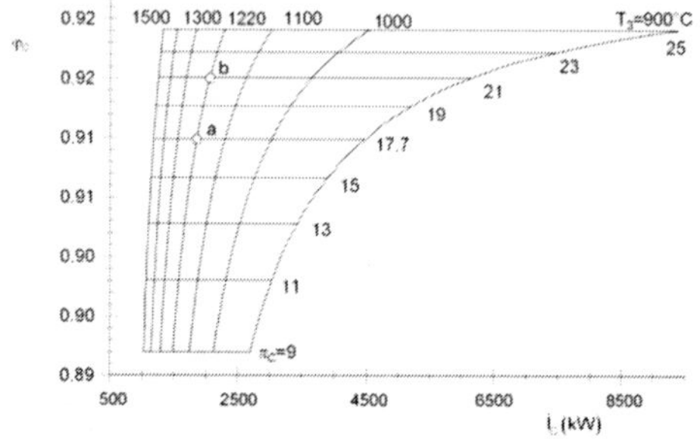

(**a**) Exergetic efficiencies of compressor *versus* irreversibility rate in the compressor for different high pressure turbine inlet temperatures and different pressure ratios of the compressor.

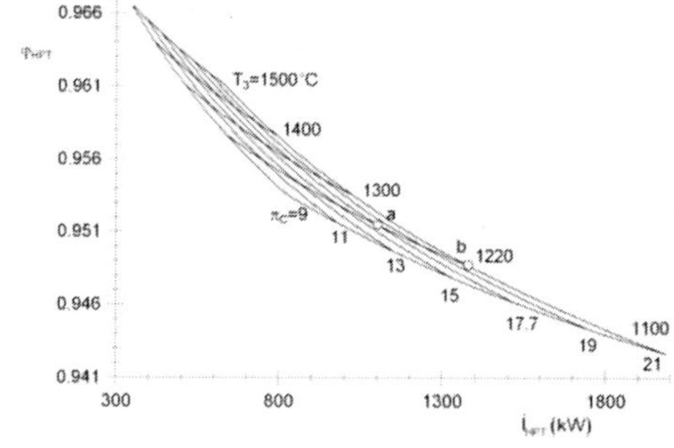

(**b**) Exergetic efficiencies of the high pressure turbine *versus* irreversibility rate of the high pressure turbine for different high pressure turbine inlet temperatures and different pressure ratios of compressor.

Figure 8. Exergetic efficiency of compressor and high pressure gas turbines on irreversibility rates for different high pressure turbine inlet temperatures and pressure ratios of the compressor.

Figure 8b shows the exergetic efficiency of high pressure turbine at different irreversibility rates in the high pressure turbine for different high pressure turbine inlet temperatures and different pressure ratios of the compressor. For a high pressure turbine inlet temperature of 1220 °C and a pressure ratio of compressor of 17.7 (point a), the exergetic efficiency high pressure turbine is 95.14%, and the irreversibility rate is 1084.26 kW. Note also that by increasing the high pressure turbine inlet temperature to 1500 °C, the exergetic efficiency is increased 1.11% and the irreversibility rate decreases to 532 kW.

The results displayed in Figure 8a and Figure 8b show that the exergetic efficiency of the high pressure turbine is higher than from the compressor, and the irreversibility rates of the high pressure turbine are lower than from the compressor.

Figure 9a shows that by increasing the high pressure turbine inlet temperature and keeping a constant pressure ratio of the compressor, the thermal efficiency increases and the irreversibility rates in the compressor decreases. Note, for instance, that for 1220 °C and a pressure ratio of 17.7, the thermal efficiency is 34% and the irreversibility rates are 1847.17 kW; decreasing the temperature to 1100 °C and keeping constant the pressure ratio, the thermal efficiency is decreased 1.8%, and irreversibility rates increase 422.3 kJ/kg.

Figure 9b shows the thermal efficiency variation as a function of the irreversibility rates in combustion chamber for different high pressure turbine inlet temperatures and pressure ratios of the compressor. It clearly shows that for a high pressure turbine inlet temperature of 1220 °C and a compression ratio of 17.7 (point a), the thermal efficiency is 33.92% and the irreversibility rate is 8490.2 kW. Increasing the compression ratio to 21 (point b) and keeping the temperature constant, the thermal efficiency is increased to 34.22% and the irreversibility decreases 4.14%. In combustion chamber, larger losses in aeroderivate gas turbine are generated.

Figure 10a,b show the thermal efficiency variation as a function of the irreversibility rates in low and high pressure turbines, respectively, for different high pressure turbine inlet temperatures and pressure ratios of the compressor. The high pressure turbine generates twice the amount of irreversibility as the low pressure turbine. Note, for instance, for a high pressure turbine inlet temperature of 1220 °C, and a decrease of the compression ratios from 21 (point b) to 17.7 (point a) there is a reduction of the irreversibility in the low pressure turbine of 100 kW and of 500 kW in high pressure turbines.

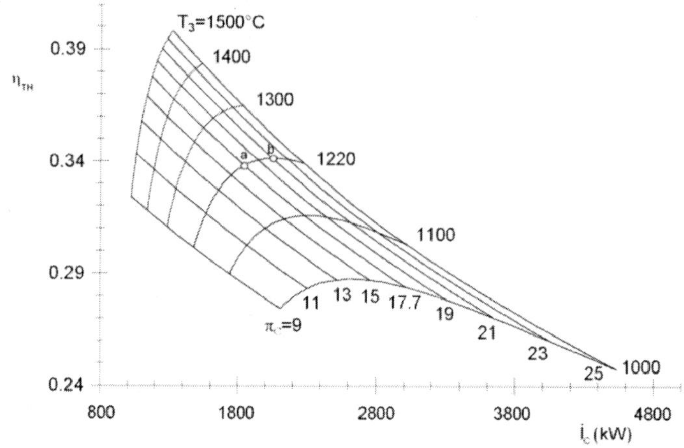

(a) Compressor.

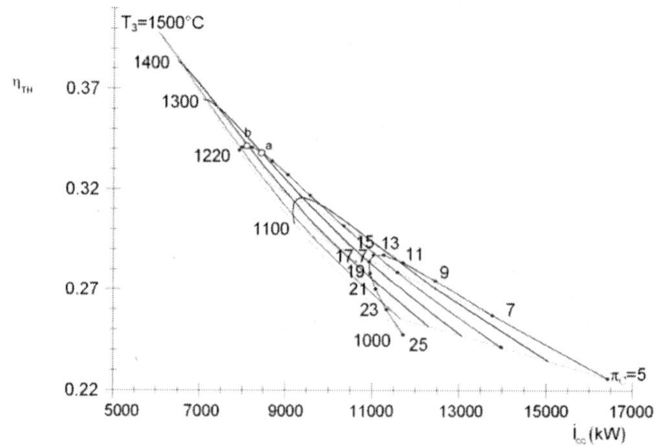

(b) Combustion chamber.

Figure 9. Thermal efficiency on the irreversibility rate for different high pressure turbine inlet temperatures and pressure ratios of the compressor.

The values of the irreversibilities agree with the tendencies of the results that present Hakan Aydin [2]; nevertheless, it disagrees with tendencies of the results of the low pressure turbine, maybe because he analyzes an areoderivate gas turbine with an NGV by increasing the low pressure turbine inlet temperature.

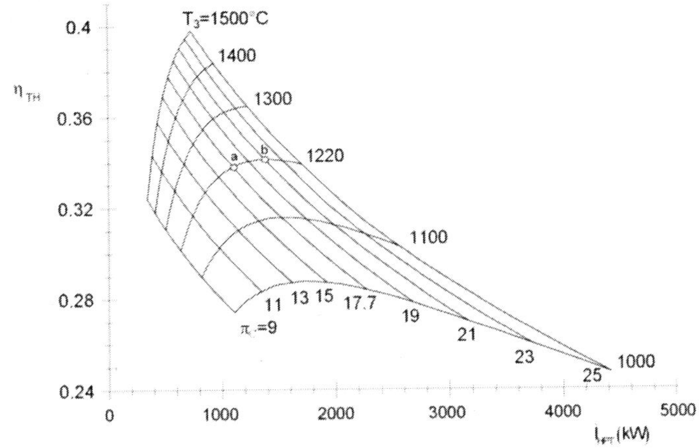

(a) High pressure turbine.

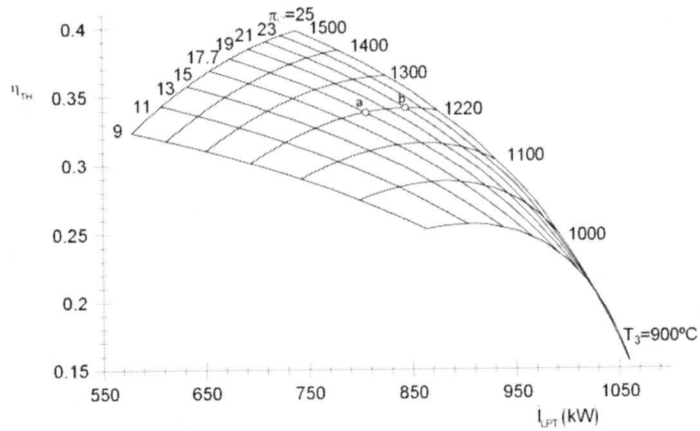

(b) Low pressure turbine.

Figure 10. Thermal efficiency on the irreversibility rate for different high pressure turbine inlet temperatures and pressure ratios of the compressor.

5. CONCLUSIONS

A simple methodology has been presented in this paper to estimate the values of the pressure ratio and the high pressure turbine inlet temperature for the operation of an aeroderivate gas turbine. The presented methodology shows that the high pressure turbine inlet temperature is 1220 °C and the pressure ratio is 17.7. The information to estimate the high pressure turbine inlet temperature and pressure ratio was obtained from operation actual values of an aeroderivate gas turbine Mars 100 that is operating on an offshore platform.

With the same methodology, a parametric analysis is pursued and the results presented show that the larger irreverisibilities are generated in the combustion chamber with 36.48% of the total inlet exergy, in the compressor shows 8.21%, the high pressure turbine is 3.05%, and low pressure turbine is 3.55%, respectively. Results show that the exergetic efficiencies are: compression of 91.7%, combustion chamber of 92.6%, low and high pressure turbines of 95% and 93.5%, respectively. The thermal efficiency for the optimal pressure ratio is 1% lower than that by the pressure ratio of 17.7; nevertheless, the irreversibilities decrease approximately 1 GW when a11.86 MW is generated, due to the fact that the irreversibilities are incremented in the process of compression and the expansion of the free turbine. The presented results also show that, if the compression efficiency decreases from 86% to 84%, the irreversibilities of the compressor increase 18.85%; the irreversibilities in the combustion chamber are increased 1.48%; the irreversibilities in the high pressure turbine are increased 6.97% and in the free turbine, 0.14%. In this manner, this paper provides evidence that if the turbine efficiency decreases, the combustion chamber is more affected due to the increase of the fuel flow rate by producing the power required.

ACKNOWLEDGEMENTS

The authors would like to thank the reviewers for their valuable comments and suggestions to improve the present work.

AUTHOR CONTRIBUTIONS

R. Lugo-Leyte designed the research; M. Salazar-Pereyra and I. Aguilar-Adaya developed the methodology and performed the simulation; H. D. Lugo Méndez and J. G. Vázquez Vargas analyzed the data; J. J. Ambriz-García and R. Lugo-Leyte wrote the paper. All authors have read and approved the final manuscript.

NOMENCLATURE

cp specific heat capacity, at constant pressure; [kJ/kgK]

$\dot{E}$ exergy rate; [kW]

g Gibbs free energy; [kJ/kg]

h specific enthalpy; [kJ/kg]

HR heat rate; [kJ/ kW h]

i specific irreversibility; [kJ/kg]

$\dot{I}$ irreversibility rate; [kW]

$\dot{m}$ mass flow rate; [kg/s]

n polytropic index; [-]

P pressure; [bar]

LHV low heating value of fuel; [kJ/kg]

q specific heat supplied or rejected; [kJ/kg]

$\dot{Q}$ heat transfer rate; [kW]

r reheat degree; [kJ/kg]

R ideal gas constant; [kJ/kg K]

s specific entropy; [kJ/kg K]

T temperature; [°C o K]

AFT adiabatic flame temperature; [°C o K]

w specific work output; [kJ/kg]

$\dot{W}$ power; [W]

x ratio of ideal gas constant to specific heat at constant pressure; [-]

y ratio of temperature inlet high pressure turbine to temperature inlet compressor; [-]

Greek letters

Δ drop, increase or difference

γ cP/cv, isentropic index; [= 1.4 air]

ε specific exergy; [kJ/kg]

η efficiency; [-]

φ exergetic efficiency; [-]

π pressure ratio; [-]

τ exergetic temperature factor; [-]

Subscripts

0 dead state

a air

C compression

cc combustion chamber

f fuel

g exhaust

in added

m mechanical

max maximum

p product

avg average

r reactive

SUM supplied

HPT high pressure turbine; [-]

LPT low pressure turbine; [-]

GT gas turbine

tp output

TH thermal

Superscripts

Qf heat transfer

REFERENCES

1. Turan, O.; Aydin, H. Exergetic and exergo-economic analyses of an aero-derivative gas turbine engine. Energy 2014, 74, 638–650.
2. Aydin, H. Exergetic sustainability analysis of LM6000 gas turbine power plant with steam cycle. Energy 2013, 57, 766–774.
3. Lugo-Leyte, R.; Zamora-Mata, J.M.; Toledo-Velázquez, M.; Salazar-Pereyra, M.; Torres-Aldaco, A. Methodology to determine the appropriate amount of excess air for the operation of a gas turbine in a wet environment. Energy 2010, 35, 550–555.
4. Nkoi, B.; Pilidis, P.; Nikolaidis, T. Performance of small-scale aero-derivative industrial gas turbines derived from helicopter engines. Propuls. Power Res. 2013, 2, 243–253.
5. Ghaebi, H.; Amidpour, M.; Karimkashi, S.; Rezayan, O. Energy, exergy and thermoeconomic analysis of a combined cooling, heating and power (CCHP) system with gas turbine prime mover. Int. J. Energy Res. 2011, 35, 697–709.
6. Bejan, A. Advanced Engineering Thermodynamics; Wiley: Hoboken, NJ, USA, 2006; pp. 204–225.
7. Kotas, T.J. The Exergy Method of Thermal Plant Analysis; Butterworths: London, UK, 1985; pp. 99–150.
8. Dincer, I.; Rosen, M.A. Exergy: Energy, Environment and Sustainable Development; Elsevier: Oxford, UK, 2007; pp. 50–75.
9. Tsatsaronis, G. Thermoeconomic analysis and optimization of energy systems. Prog. Energy Combust. Sci. 1993, 3, 227–257.
10. Khaliq, A. Exergy analysis of gas turbine trigeneration system for combined production of power heat and refrigeration. Int. J. Refrig. 2009, 3, 534–545.
11. De Sa, A.; Zubaidy, S.A. Gas turbine performance at varying ambient temperature. Appl. Therm. Eng. 2011, 31, 2735–2739.
12. Najjar, Y.S.; Al-Absi, S. Exergy analysis for greener gas turbine engine arrangements. J. Eng. Thermophys. 2013, 22, 247–256.
13. Tona, C.; Raviolo, P.A.; Pellegrini, L.F.; de Oliveira Júnior, S. Exergy and thermoeconomic analysis of a turbofan engine during a typical commercial flight. Energy 2010, 2, 952–959.

14. Agudelo, A.; Valero, A.; Torres, C. Allocation of waste cost in thermoeconomic analysis. Energy 2012, 45, 634–643.

15. Sue, D.-C.; Chuang, C.-C. Engineering design and exergy analyses for combustion gas turbine based power generation system. Energy 2004, 29, 1183–1205.

16. Bilgen, E. Exergetic and engineering analyses of gas turbine based cogeneration systems. Energy 2000, 25, 1215–1229.

17. Facchini, B.; Fiaschi, D.; Manfrida, G. Exergy Analysis of Combined Cycles Using Latest Generation Gas Turbines. J. Eng. Gas Turbines Power 2000, 122, 233–238.

4

CHAPTER

STUDYING THE ROLE PLAYED BY EVAPORATIVE COOLER ON THE PERFORMANCE OF GE GAS TURBINE EXISTED IN SHUAIBA NORTH ELECTRIC GENERATOR POWER PLANT

O. R. AL-Hamdan, A. A. Saker

Water Resources Dept., High Institute of Energy, Public Authority of Applied Education & Training, Kuwait

ABSTRACT

It's well known that the performance of a gas turbine (efficiency, heat rate and power generated) is largely dependent on mass flow rate of air, inlet air temperature and turbine inlet temperature (TIT). As turbine inlet temperature is dependent on quantity of burned fuel so that this factor is dropped out from this paper. It's also known that gas turbines are constant volume machines i.e. at a given shaft speed they always move the same volume of air, but the power output of a turbine depends on the flow of mass through it. This is precisely the reason why on hot days, when air is less dense, power output falls off. A rise of one degree Centigrade temperature of inlet air decreases the power output by 1% and at the same time heat rate of the turbine also goes up. This is a matter of great concern to power producers. Many techniques have been developed to cool the inlet air to gas turbine. Some of these techniques to decrease the inlet air temperature are discussed here. The evaporative cooling technique is taken as a case study in this paper. A comparative studying is carried out between a unit using this technique and the same unit when the evaporative cooler is idle. The results advert to an increase in power output by 11.07% and a decrease in heat rate by approximately 4% when inlet air temperature drops from 50°C to 26°C.

Keywords: Gas Turbine; Shaft Speed; Power Output; The Evaporative Cooling Technique

1. INTRODUCTION

Electric generator power plants play a great role in developing countries. The modern techniques in generating electricity from fossil fuel are by using the so-called combined units through combined cycle. The combined cycle in Shuaiba North power plant consists of three unit gas turbine and one unit steam turbine. The performance characteristics of these units are the most important judging element. The main factors affecting unit performance characteristics are the inlet air temperature to the compressors of the gas turbine units. The research will focus on the system of inlet air cooling to a compressor of GE Gas turbine unit, 250 MW capacities as a case study.

2. LITERATURE SURVEY

An inlet cooling system is a beneficial option for applications where significant operations occur during times of warmer temperatures combined with low relative humidity. Cooler air is denser, resulting in a higher mass flow rate to the compressor. This results in an increase in turbine output and efficiency [1].

Evaporation of water is one of the simplest and oldest methods of cooling air. Even with the sophisticated technology available today, including mechanical chillers, absorption chillers and thermal energy storage systems, evaporative cooling remains a most cost-efficient method for temperature control of the gas turbine inlet air supply [2].

Evaporative cooling can be achieved by many methods. In practical two forms are typically used. Media type evaporative cooling and spray type evaporative cooling.

The most famous for media type evaporative coolers consist of re-circulated water sprayed over an extended surface media mounted downstream from the inlet air filters. As inlet air passes through the water soaked evaporative cooler media, evaporation occurs performing a double function:

1) Energy in the form of heat is removed from the air and in the same time that energy is used to evaporate water in the media; and 2) Water vapor content of the air increases due to evaporated water approaching saturation along constant wet bulb lines.

These two processes increase the density of the air, which in turn increases the mass flow and output of the gas turbine.

3. CYCLE THERMODYNAMIC CONSIDERATION

3.1. Gas Turbine Cycle Thermodynamics

The compression process consumes as much as 66% of the total work produced by the gas turbine and therefore any means of reducing the work of compression will enhance the power output of the gas turbine. The compressor work per unit mass of air is given by:

$$\frac{W}{m} = \frac{h_2' - h_1}{\eta_c} = \frac{C_p\left(T_2' - T_1\right)}{\eta_c} = \frac{C_p T_1\left[\left(\frac{P_2}{P_1}\right)^{\frac{\gamma-1}{\gamma}} - 1\right]}{\eta_c} \quad (1)$$

In examining this equation, it can be seen that increasing T_1 would increase the compression work.

Assuming no pressure losses, and equal specific heats in the cycle gas turbine, thermal efficiency can be reduced to the equation below (Sorenson, 1950) [3].

$$\eta_{th} = \frac{\dfrac{\eta_t T_3}{B} - \dfrac{T_1}{\eta_c}}{\dfrac{T_3 - T_1}{B-1} - \dfrac{T_1}{\eta_c}} \quad (2)$$

where:

$$B = \left[\left(\frac{P_2}{P_1} \right)^{\frac{\gamma - 1}{\gamma}} \right]$$

η_c = Compressor efficiency, η_t = Turbine efficiency, T_3 = Turbine inlet temp., P_2, P_1 = Cycle pressure limits.

Examination of this equation shows that the cycle efficiency decreases with an increase in compression inlet temperature. When the inlet temperature goes up the compressor discharge pressure and temperature drops and more fuel is required to attain the same turbine inlet temperature (TIT).

An increase in ambient temperature T_1 causes the numerator, representing net power output to decrease at a faster rate than the denominator, representing the heat delivered. So the cycle efficiency decreases with increasing inlet compressor temperature.

Figure 1 shows a T-S diagram on a hot day the drop in pressure ratio that occurs on the hot day can be seen on this diagram. The cycle peak temperature is still limited as before and so the expansion ratio also drops meaning that less work is extracted from the turbine. As the compressor work increases and the turbine work decreases, the output power drops [3].

Further, as the ambient temperature increases, the pressure ratio of the compressor decreases and more fuel has to be added to attain the same turbine inlet temperature (TIT) this further decreases the output of the gas turbine unit and the cycle efficiency drops [3].

3.2. Effect of Inlet Air Temp. on GE Gas Turbine Performance Characteristics

GE offers a table explains the effect of inlet air temperature on the power output and heat rate of many models of gas turbine. Table 1 provides a complete listing of the available outputs and heat rates of the GE heavy-duty gas turbines [4].

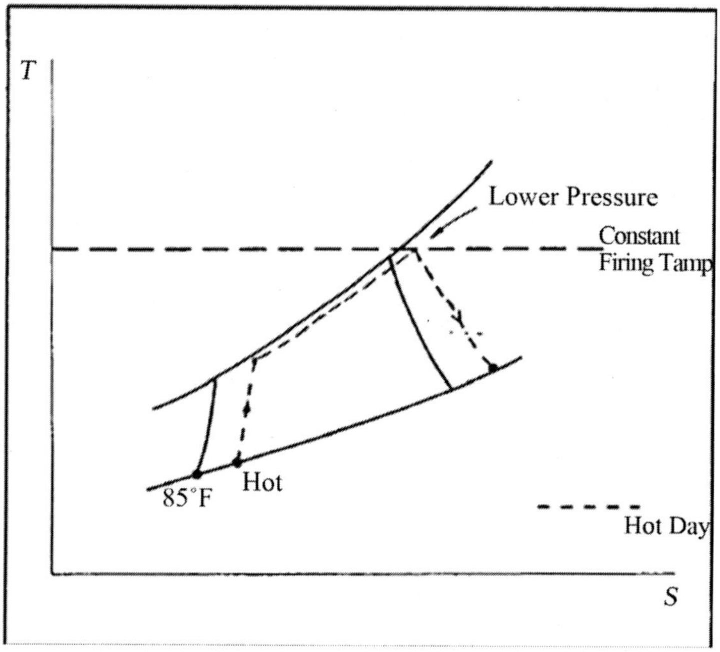

Figure 1. T-S Diagram of Simple Brayton Cycle on Hot Day.

Table 1. GE gas turbine performance characteristics— Mechanical drive gas turbine ratings.

Model	ISO rating continuous (KW)	Heat rate (KJ/KWh)	Mass flow (Kg/s)	Exhaust temp. (degree °C)
M3142 (J)	11.290	13.440	53	542
M3142R (J)	10830	10.450	53	370
M5261 (RA)	19.690	13.270	92	531
M5322R (B)	23.870	10.000	114	352
M5352 (B)	26.110	12.490	123	491
M5352R (C)	26.550	9.890	121	367
M5382 (C)	28.340	12.310	126	515
M6581 (B)	38.290	11.060	134	545

3.3. Factors Affecting Performance of Gas Turbine

The performance of Gas Turbine viz. the power output and the heat rate (measure of efficiency, i.e. the amount of energy consumed per kWh of electricity produced) depends on the following major parameters [5].

1) Site altitude i.e. atmospheric pressure.
2) Inlet pressure drops in the filters and intake system.
3) Outlet pressure drops in the HRSG (Heat Recovery Steam Generator).
4) Site design temperature.
5) Site design relative humidity corresponding to the site design temperature.

3.4. Impact of Higher Inlet Air Temperature on Gas Turbine Performance

A gas turbine is a constant volume machine i.e. the volume of air compressed is fixed, irrespective of ambient temperature. Hence, as the temperature of air raises the density of air decreases and the mass flow rate of compressed air gets reduced. The power output of the gas turbine is proportional to the mass flow rate of air. Thus as the ambient temperature increases, the power output decreases. Further, the efficiency of the gas turbine also falls, as more power is required to compress warmer air, from the above paragraph (factors affecting performance of gas turbine). For a given site and the configuration, the first three parameters are constant and cannot be changed. However, it is possible to alter the other two parameters and obtain a higher output and improved efficiency by cooling the air before it is admitted in the gas turbine compressor section. Impact of higher ambient air temperature is very well spelled in a given Figure 2 [5], and Table 2 [6].

4. METHODS CURRENTLY ADOPTED BY INDUSTRY

Industry has implemented various techniques to reduce ambient temp. They are as follows [5]:

1) Vapor Compression:

Various refrigerant type air-chilling systems are available in the market. The biggest advantage of this is that one can reduce temp of inlet air up to

15°C. The biggest disadvantage in this system is higher power consumption to cool inlet air and pressure drop in the air, thus adversely affecting the performance of compressor.

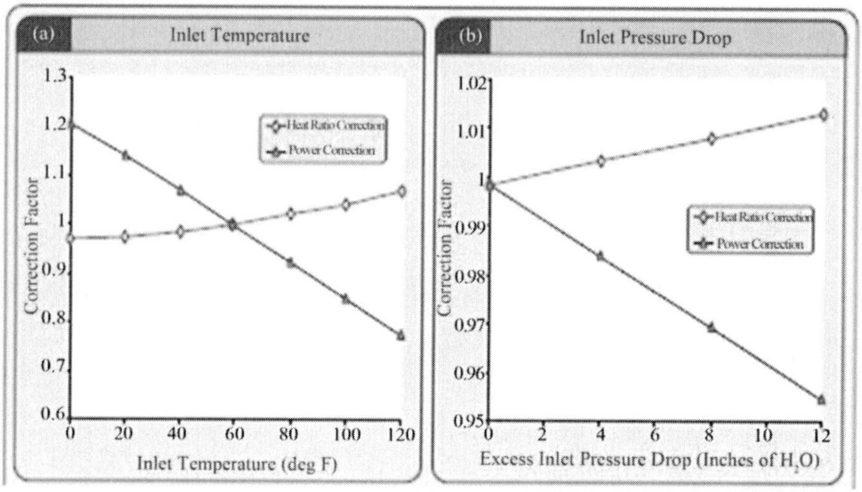

Figure 2. Changes in Power Output and Heat Rate for a 40 MW Gas Turbine Output to: (a) Inlet Air Temperature. (b) Inlet Pressure Drop.

Table 2. Performance of various gas turbine models with respect to inlet air temperature.

Model		GT 10 B			Frame 5 (5371 PA)			Frame 6 (6561 B)			*LM 6000PC		
Inlet temp.	°C	15	35	40	15	35	40	15	35	40	15	35	40
GT power	MW	23.1	20.1	18.9	25.2	21.4	20.4	38.6	33.5	32.3	40.9	31.9	27.4
GTG efficiency	%	33	32.11	31.8	28	26.4	25.9	31.5	30.3	30	40.6	37.8	35.4
Decrease in power w.r.t 15°C	%	0	13	18	0	15.1	19.2	0	13.1	16.3	0	22	32.9
Decrease in efficiency w.r.t 15°C	%	0	3	4	0	5.82	7.64	0	3.8	5.04	0	7.04	12.9

2) Vapor Absorption Chillers:

Steam is used to give the chilling effect and one can achieve 15°C inlet air temperature.

This system consumes steam which reduces output of the steam turbine .It also reduces the inlet air pressure which adversely affects the performance of compressor.

3) High Pressure Fog system:

It's one of the recent technologies employed for inlet air cooling. It's similar to evaporative cooling, but instead of using water as an evaporative medium, the water is atomized into billions of supper small droplets thereby creating a large evaporative surface area. With fog overcooling arrangement, one can generate more power. The biggest disadvantages of this technique are high cost of high pressure nozzles, high vibration, prices of high pressure pumps and the ability of erosion in compressor blades.

4) Evaporative Cooler:

Evaporative cooling works on the principle of reduceing the temperature of an air stream through water evaporation. The process of converting water from liquid to a vapor state requires energy. This energy is drawn from the air stream, the result being cooler and more humid air. The effectiveness of an evaporative cooling system depends on the surface area of the water exposed to the air stream and the residence time. Performance of the system is restricted by the amount of moisture present in the air. It works well in low humidity area.

5. THE PRESENT WORK (CASE STUDY)

For better choice between the cooling methods it's necessary to take into consideration the following factors:

1) Capital cost.

2) Maximum achievable cooling range.

3) Auxiliary power/steam consumption.

4) Reliability of the system.

5.1. Evaporative Cooler in Shuaiba North Gas Turbine Power Plant

The system existed in Shuaiba North gas turbine power plant for decreasing the inlet air temperature to the compressor consists of two centrifugal water pumps with capacity 20 m^3/hr and 120 m head. The two pumps are parallel arrangement. Figure 3 shows the line diagram of the system arrangement.

The evaporative cooling system arrangement is also shown in Figure 4 [7].

5.2. Thermodynamic Relationships of Evaporative Cooler System

The function of the evaporative cooler is to increase turbine output by lowering the inlet dry bulb temperature and increasing the air density through the evaporation of water into the inlet airflow. Re-circulated water introduced over the top of the evaporative cooler media drains through and wets the media. Filtered ambient air comes into contact with the wetted media where the air is cooled by the heat of evaporation of water. Gas turbine mass flow rate increases due to the lower air temperature and increased water content. Higher mass flow results in increased power output from the gas turbine. When discussing media type evaporative coolers, a couple of key parameters are essential to the proper operation and maintenance of evaporative coolers. These evaporative cooler parameters are: Saturation Efficiency, Evaporation Rate, Blow down Rate, Makeup Water Rate, Water Carryover, and Water Bypass

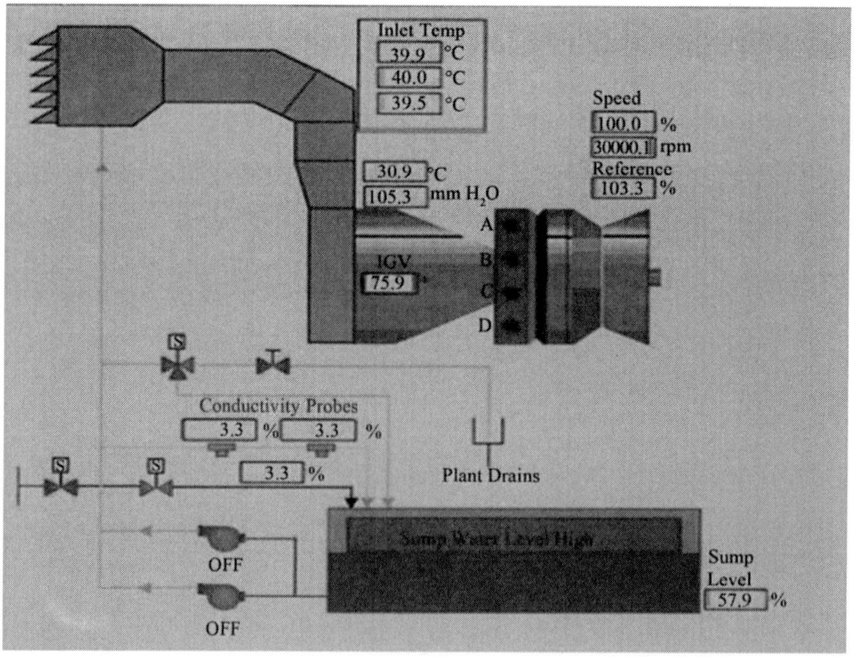

Figure 3. Schematic Diagram of Evaporative Cooling Components.

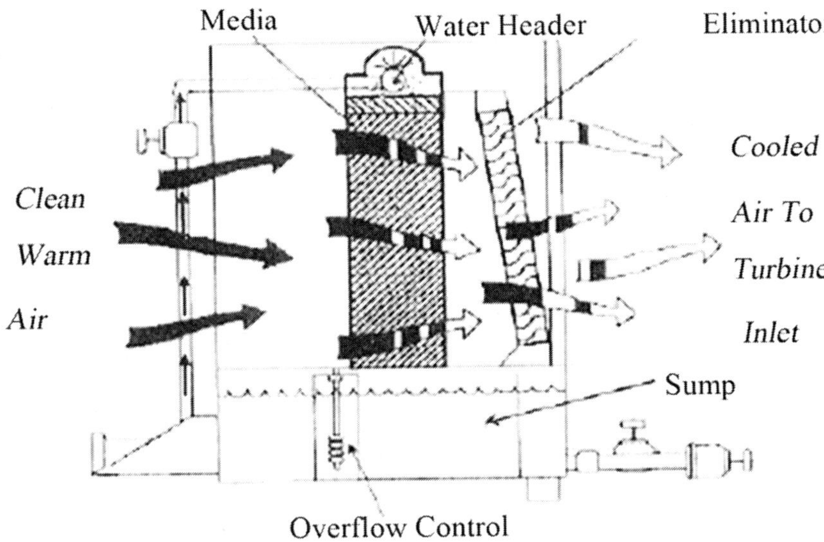

Figure 4. Schematic Diagram of an Evaporative Cooling.

5.2.1. Saturation Efficiency (Effectiveness)

The performance of an evaporative cooler is based on the ratio of the number of degrees it can cool the air compared to the wet bulb temperature depression. As air passes through an evaporative cooler, the dry bulb temperature is lowered, but the wet bulb temperature remains approximately the same.

The dry bulb temperature will then approach the wet bulb temperature as the air is cooled and the relative humidity will rise (due to moisture added to the airflow through evaporation). The entering wet bulb temperature is a function of the entering conditions of the airflow (air dry bulb temperature, ambient pressure or altitude with respect to sea water level, and relative humidity). Since an evaporative cooler operates along constant wet bulb temperature lines, the entering wet bulb temperature determines the maximum amount of cooling that can be achieved. We can define the saturation efficiency of an evaporative cooler as the ratio between the actual amounts of cooling achieved (difference between entering and exiting airflow temperatures) and the maximum allowable cooling per the entering conditions (Wet Bulb Depression). The effectiveness of an evaporative cooler may be calculated using the following formula and Figure 5 [7].

$$\eta = \frac{T_{DBE} - T_{DBL}}{WBD} = \frac{T_{DBE} - T_{DBL}}{T_{DBE} - T_{WBE}} \tag{3}$$

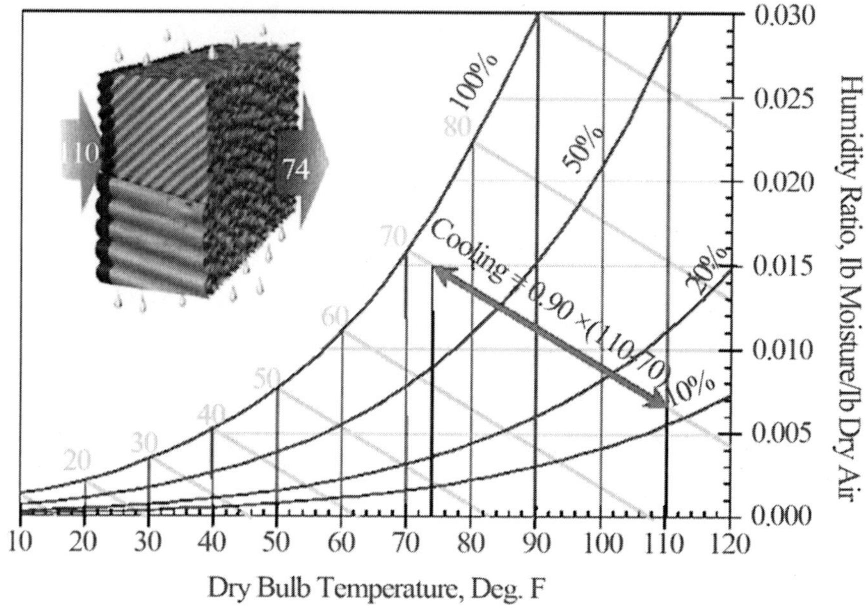

Figure 5. Evaporative Cooler Process on Psychometric Chart.

where:

η = Efficiency in percent.

T_{DBE} = Entering Dry Bulb Temp T_{WBE} = Entering Wet Bulb Temp T_{DBL} = Leaving Dry Bulb Temp WBD = Wet Bulb Depression, or $= T_{DBE} - T_{WBE}$

1) Dry Bulb Temperature The temperature as measured by a standard thermometer.

2) Wet Bulb Temperature If the air is at a temperature above its dew point, evaporation of water will occur in the wick causing cooling and the reading of a temperature below the dry bulb temperature, the wet bulb and dry bulb temperatures are equal at 100% humidity.

The Wet Bulb Depression is the difference between the ambient dry bulb and the wet bulb temperature. The Wet Bulb Depression is the maximum amount of cooling achievable with a 100% efficient evaporative cooler system.

3) Humidity Humidity, which is expressed by the water vapor pressure in air, is controlled by the ambient temperature. Relative Humidity is the ratio of the vapor pressure of water in air compared the saturated water vapor pressure at the same temperature. Dew Point Temperature of the air is the temperature at which condensation of liquid water may occur.

4) Ambient Pressure The temperature and barometric pressure of atmospheric air vary considerably with altitude as well as with local geographic and weather conditions. The standard atmosphere gives a standard of reference for estimating properties of air at various altitudes. At sea level, the standard barometric pressure is 14.696 Ib/in^2 [29.921 in. Hg] and the standard temperature is 59°F [15°C]. The temperature is assumed to decrease linearly with increasing altitude while the specific volume increases. The lower atmosphere is assumed to consist of dry air that behaves as a perfect gas.

5.2.2. Evaporation Rate

The evaporation rate is the amount of water that is added to the gas turbine airflow as a result of the energy transfer between airflow and water, and the subsequent evaporation of water into the air. The amount of water evaporated into the air or evaporation rate will depend on the entering ambient air conditions. Factors such as dry bulb temperature, relative humidity, and air ambient pressure determine the maximum amount of water that may be evaporated before air becomes "saturated" with water.

The amount of water evaporated in gallons per minute as the air passes through the cooler can be calculated using the psychometric chart using the following formula:

$$\text{Evaporation} = \frac{V_{\text{air}}\left(W_L - W_E\right)}{\rho_{\text{water}} v_{\text{air}}} = \frac{\rho_{\text{air}} V_{\text{air}}\left(W_L - W_E\right)}{\rho_{\text{water}}} = \frac{m_{\text{air}}\left(W_L - W_E\right)}{\rho_{\text{water}}} \tag{4}$$

where:

V_{air} = Actual volumetric flow rate of air (cubic feet of air per minute or CFM)

ρ_{water} = Density of water (lb water/gal) at T_{WBE}

ρ_{air} = Density of air (lb air/ft^3) at entering airflow conditions W_E = Moisture content of entering air (lbs. water/lb. dry air)

W_L = Moisture content of leaving air (lbs. of water/lb. dry air)

m_{air} = Mass airflow rate (lb. dry air/min) at entering conditions V_{air} = Specific volume of air (ft^3/lb. dry air) at entering conditions Similarly, since the evaporation rate is a function of the temperature of the ambient air and airflow, the amount of water evaporated may be obtained from the energy balance equation between air entering and leaving the evaporative cooler media:

$$\text{Evaporation} = \frac{m_{\text{air}} C_{p_{\text{air}}} \left(T_{DBE} - T_{DBL}\right)}{\rho_{\text{water}} h_{fg\,\text{water}}} = \frac{\rho_{\text{air}} V_{\text{air}} C_{p_{\text{air}}} \eta \left(T_{DBE} - T_{WBE}\right)}{\rho_{\text{water}} h_{fg\,\text{water}}} \quad (5)$$

where:

T_{DBL} = Dry bulb temperature at leaving from evaporative cooler.

T_{DBE} = Dry bulb temperature at entrance to evaporative cooler.

T_{WBE} = Wet bulb temperature at entrance to evaporative cooler.

C_{Pair} = Specific heat of air at T_{DBE} (Btu/lb. °F)

$h_{fgwater}$ = Heat of evaporation of water (Btu/lb water) at T_{WBE}

For practical purposes, the following version of the energy balance equation is used:

$$\text{Evaporation} = \frac{V_{\text{air}} \left(T_{DBL} - T_{DBE}\right)}{500000} = \frac{V_{\text{air}} \eta \left(T_{DBE} - T_{WBE}\right)}{500000} \quad (6)$$

5.2.3. Blow down (Bleed-Off) Rate

Water always contains a certain amount of dissolved minerals. The process of evaporative cooling removes pure water as vapor from the re-circulating flow and leaves behind the solids that had been dissolved in the water when it was added as makeup. Accordingly, enough water must be blown down from the re-circulating flow to control the level of these solids and to avoid build-up of insoluble minerals on the pad surface (commonly referred to as "scaling"), which results in an increase in pressure drop, and a loss of evaporation area and efficiency.

Blow down is the flow of water that must be continuously removed from the cooler in order to maintain the chemistry of the re-circulating water at the design value. It is a function of evaporation rate and the cycles of concentration that can be achieved with a given water quality in the system.

$$\text{Blow down} = \text{Evaporation rate}/(\text{No. of cycle-1}) \tag{7}$$

The chemistry of the sump is established by determining the maximum cycles of concentration that the makeup water can go through. By definition, the number of cycles of concentration is equal to the number of times that incoming (makeup) water may be re-circulated before it is removed (bled) from the system. Two methods are commonly used for controlling blow down and maintaining the desired chemistry in the sump of the evaporative cooler: Constant flow and Conductivity control blow down.

5.2.4. Makeup Water Rate

Makeup water is the water added to the sump of the evaporative cooler to replace the water lost by evaporation into the airflow and the water removed from the sump through blow down or bleed-off as required.

$$\text{Makeup Rate}(\text{gpm}) = \text{Evaporation Rate}(\text{gpm}) + \text{Blow down Rate}(\text{gpm}) \tag{8}$$

When conductivity control is used to control blow down, the makeup water required while blow down is on will be higher than the average identified by the makeup rate formula. When the blow down is off (Blow down rate = 0 gpm), the required make up rate will be equal to the current evaporation rate.

A high makeup water rate is typically required during startup of the evaporative cooler system in order to ensure enough water is available out of the sump (recirculation) tank during initial wetting of media.

5.3. Role Played by Evaporative Cooler on *Power* Output and, Heat Rate of Shuaiba North Gas Turbine Power Plant

From Table 3 which show the properties of air at entering and leaving from evaporative cooler, When using the evaporative cooling techniques the power output = 224.9 MW and heat rate = 9882.2 kj/kw hr - when the ambient conditions are: dry bulb inlet temperature T_{DB} = 42.48 C, inlet pressure = 0.99 bar and relative humidity = 12%, the leaving conditions from the evaporative cooling are as follows:

Exit dry bulb temperature T_{DBL} = 22.43 CExit pressure= 0.98 bar, and the relative humidity = 85.11%.

At entrance to evaporative cooler from Figure 5.

$$W_E = 0.0054 \text{ Ib/Ib dry air.}$$

At leaving from evaporative cooler from Figure(5)

$$W_L = 0.015 \text{ Ib/Ib dry air.}$$

$$\because m_{air} = 2141.2 \text{ Ton/hr.}$$

$$\because m_W = m_a (W_L - W_E)$$

$$\therefore m_W = \frac{2414.2 \times 1000}{3600} \times 2.204 \times (0.015 - 0.0054)$$

$$= 12.58 \text{ Kg/s} \tag{9}$$

6. CALCULATION OF EVAPORATIVE COOLING EFFICIENCY

Using the data in Table 4 which shows the humidity ratio at entering and leaving from evaporative cooler:

$$\eta_{evapo.cooler} = \frac{T_{DBE} - T_{DBL}}{W_{BD}} = \frac{42.48 - 22.43}{42.48 - 2.1} = 93.34\%$$

Table 3. Properties of air at entering and leaving from evaporative cooler.

State	Entering to cooling evaporator	Leaving from cooling evaporator
Dry bulb temperature T_{DB}	42.48	22.43
Wet bulb temperature T_{WB}	21	21
Relative humidity %	12	85.11

6.1. Humidity

Amount of air entered when power output = 224.9 MW, m_{air} = 2141.2 Ton/hr.

Amount of water used when power output = 224.9 Mw is

$$m_w = m_{air} \times (W_L - W_E) = 12.58 \text{ kg/s}$$

6.2. Power Saved at Various Inlet Temperatures

Table 5 determines the amount of power saved as a result of using evaporative cooler in services in Shuaiba North gas turbine power plant at different inlet air temperature.

7. DIFFERENCES IN HEAT RATE WHEN USING THE EVAPORATIVE COOLING TECHNIQUES

Table 6 below shows the quantity of gas oil and natural gas fuels consumed in producing 225 Mw/unit from Shuaiba North gas turbine power plant when evaporative cooler in service and when it was out of services. It's clear that the quantity of fuel consumed increases when evaporative cooler is out of services, especially during hot weathers.

8. MONEY SAVED WHEN USING THE EVAPORATIVE COOLER TECHNIQUES

The unit price of gas oil and natural gas fuels in 23 January 23, 2012 are as follows [8].

Table 4. Humidity Ratio at Entering and Leaving from Evaporative Cooler.

State	Entering	Leaving	Entering temp. °C
Absolute humidity W (Ib/Ib dry air)	0.0054	0.015	42.48
Absolute humidity W (kg/kg dry air)	2.45×10^{-3}	6.8×10^{-3}	

Table 5. Output power saved as a result of using an evaporative cooler.

Output power Mw when		Power saved in Mw	Inlet air temperature
Evaporative cooler in service	Evaporative cooler idle		
215	190	25	50
220	196	24	44
224	198	26	40
225	199	26	35
225	204	21	30
225	211	14	25
225	220	5	20

Table 6. Quantity of fuel consumed at different inlet air temperature.

Power output MW	Inlet air temperature °C		Quantity of gas oil kg/s when		Quantity of natural gas kg/s when	
	Before cooler	After cooler	Evaporative cooler in service	Evaporative cooler idle	Evaporative cooler in service	Evaporative cooler idle
225	50	26	12.21	13.38	12.2	14.91
225	44	23	11.75	13.1	11.5	14.3
225	40	39.9	12.85	12.85	14	14
225	42.48	22.43	11.5	12.93	11.2	13
225	42.5	23	11.62	12.93	11.2	13
225	42.1	22.42	11.75	12.89	11.2	13

Table 7. Money saved due to using evaporative cooler in Shuaiba North gas turbine power plant.

Power output MW	Inlet air tem. °C		Quantity of gas oil (kg/s) saved due to operation of evaporative cooler	Unit price $/kg	Total Money saved from gas oil $/s	Total money saved in 4 month
	Before cooler	After cooler				
225	50	26	1.17	1.1911	1.3935	14,442,624.00
225	44	23	1.35		1.608	16,671,744.00
225	40	39.9	0		0	0
225	42.48	22.43	1.43		1.703	17,656,704.00
225	42.5	23	1.31		1.5603	16,174,080.00
225	42.1	22.42	1.14		1.3578	14,077,670.40

Table 8. Decrease in heat rate in (kj/kw·hr) When Evaporative Cooling in Service.

Power output MW	Inlet air temp. °C		Quantity of gas oil kg/s consumed when		C.V of gas oil kj/kg	Heat rate (kj/kw·hr) when		Decrease in heat rate (kj/kw·hr)	
	Before cooler	After cooler	Evaporative cooler in service	Evaporative cooler idle		Evaporative cooler in service	Cooling evaporation idle	(kj/kw·hr)	%
225	50	26	12.85	13.38	40000	8224	8563.2	339.2	3.96
225	44	23	12.63	13.1		8083.2	8384	300.8	3.59
225	40	39.9	12.85	12.85		8224	8224	0	0
225	42.48	22.43	12.49	12.93		7993.6	8275.2	281.6	3.40
225	42.5	23	12.52	12.93		8012.8	8275.2	226.4	2.74
225	42.1	22.42	12.34	12.89		7997.6	8249.6	352	4.27

Diesel (Gas oil) $3.945 per Gallon (US).

Natural Gas $1.0324 per Thermo.

When using evaporative cooler during hot weathers a lot of money can be saved as a result of decreasing the amount of fuel burned. Table 7 determine a sample of money saved when power output = 225 MW. When inlet air temperature ranging from 40°C - 50°C.

Heat Rate

Another important factor in determining the role played of evaporative cooler in Shuaiba North gas turbine power plant is the heat rate in (Kj/Kw·hr). The decrease of heat rate means the more efficient the gas turbine unit.

Table 8 determines the heat rate for Shuaiba North gas turbine power plant when power output 225 Mw. and inlet air temperature range 40°C - 50°C.

9. DISCUSSIONS

It's clear from Figure 6 that as the inlet air temperature decreases, the output power increases, which is due to increase in inlet air density which retails increases of inlet mass flow rate which leads to increases of gas turbine power output.

From Figure 7 there is a considerable amount of money saved as a result of using evaporative cooling techniques because less fuel consumed at constant turbine inlet temperature (TIT = constant).

From Figure 8 and Figure 9 as the inlet air temperature decreases, a significant decreases in heat rate is also detected.

From Figure 10 it can be noticed that there is a significant needs to use evaporative cooler especially at high temperature and low humidity.

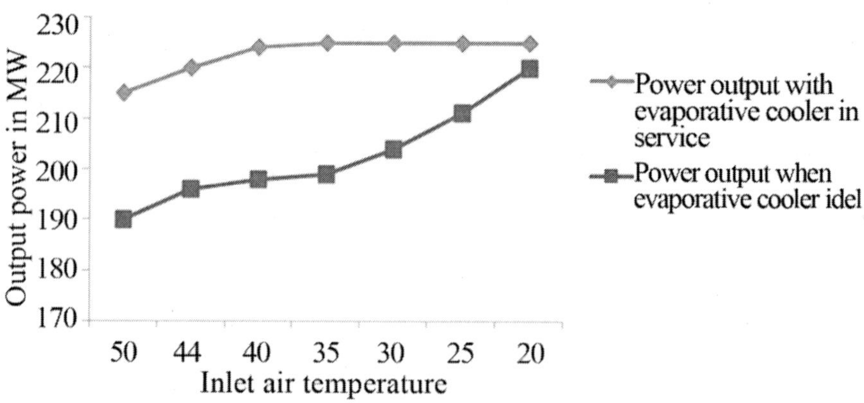

Figure 6. Effect of Inlet Air Tem. On the Power Output from Shuaiba North Gas Turbine.

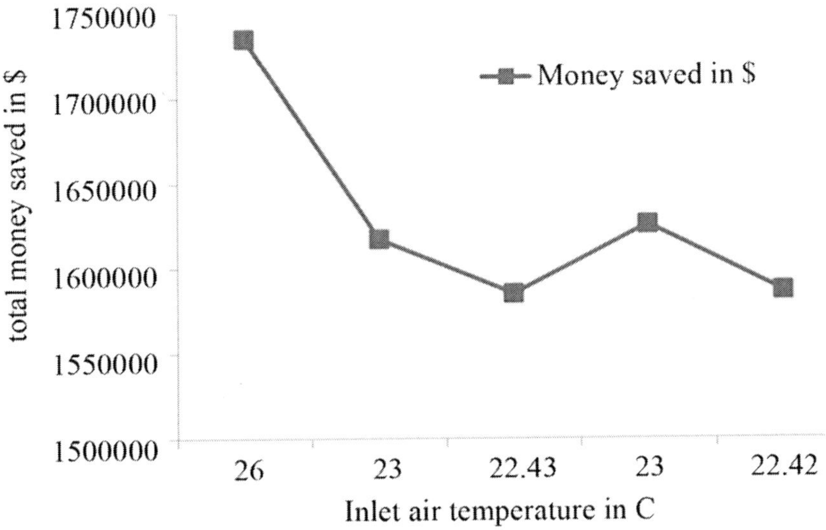

Figure 7. Money Saved When Using Evaporative Cooler in Shuaiba North Gas Turbine in $.

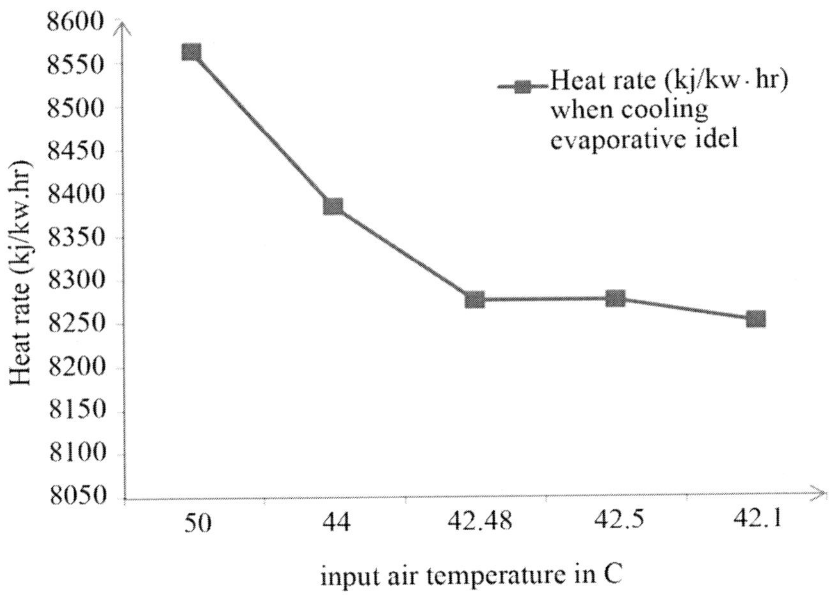

Figure 8. Effect of Input Air Tem. On Heat Rate (Kj/Kw·hr) When Cooling Evaporative Idle.

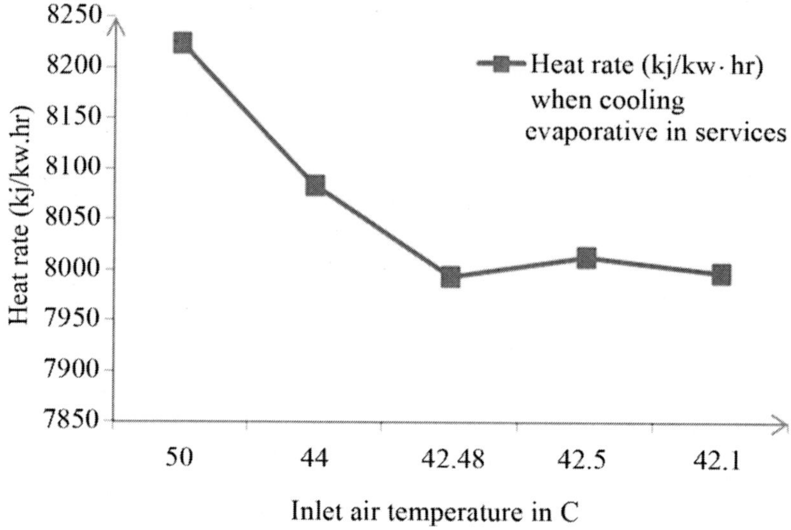

Figure 9. Effect of Input Air Tem. On Heat Rate (Kj/Kw·hr) When Cooling Evaporative in Service.

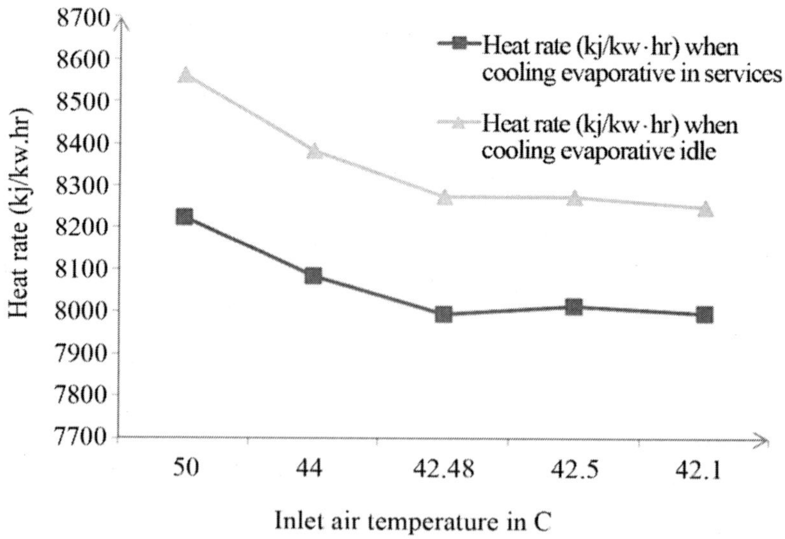

Figure 10. Comparison between Heat Rate in (Kj/Kw·hr) Versus Inlet Air Temp.

10. SUGGESTIONS AND RECOMMENDATIONS

For more power output and lower heat rate from a gas turbine, a combination of two methods of air cooling techniques can be used but, this techniques needs more analysis in future works.

REFERENCES

1. M. Chaker, C. B. M. Homji and Thomas Mee III, "Inlet Fogging of Gas Turbine Engines-Part A: Fog Droplet, Thermodynamics, Heat Transfere and Practical Considerations," Proceedings of ASME Turbo Expo 2002, 3-6 June 2002, Amsterdam.
2. M. Chaker, C. B. M. Homji and Thomas Mee III, "Inlet Fogging of Gas Turbine Engines," Proceedings of ASME Turbo Expo 2001, 4-7 June 2001, New Orleans.
3. C. B. M. Homji and Thomas Mee III, "Inlet Fogging of Gas Turbine Engines—Part B: Practical Considerations, Control, and O&M Aspects," Proceedings of ASME Turbo Expo 2000, 8-11 May 2000, Munich.
4. F. J. Brooks, "GE Gas Turbine Performance Characteristics," GE Power Systems, Schenectady.
5. P. K. Patel, "Better Power Generation from Gas Turbine along With Improved Heat Rate," 2003.
6. J. Petek and P. Hamilton, "Performance Monitoring of Gas Turbine," GE Energy, Vol. 25, No. 1, 2005.
7. Shuaiba North Power Plant Operation Manual, "Evaporative Cooler Operation & Maintenance".
8. http://oilprice.com/Energy/Oil-Prices/Natural-Gas-Analysis-for-the-Week-of-January-23-2012.html

5

CHAPTER

INTEGRATED COMPUTATIONAL MATERIALS ENGINEERING FROM A GAS TURBINE ENGINE PERSPECTIVE

Ann Bolcavage[1], Paul D Brown[2], Robert Cedoz[1], Nate Cooper[1], Chris Deaton[1], Daniel R Hartman[1], Akin Keskin[2], Kong Ma1, John F Matlik[1]*, Girish Modgil[1] and Jeffrey D Stillinger[1]

(1)Rolls-Royce Corporation, Indianapolis

(2)Rolls-Royce PLC

ABSTRACT

In 2008, the National Research Council published a landmark report on Integrated Computational Materials Engineering (ICME) and defined it as 'an emerging discipline that aims to integrate computational materials science tools into a holistic system that can accelerate materials development, transform the engineering design optimization process, & unify design and manufacturing'. ICME is becoming a critical enabler for reducing the design/make cycle time and getting complex systems into production more quickly. There are several reasons why this is the case. Firstly, ICME allows materials experts to develop new material systems and methods of manufacture much more quickly. Advanced new materials and their associated manufacturing processes can be tailored to deliver products that meet design requirements quickly and more effectively in terms of cost and performance. Secondly, ICME enables design processes to quantify cause and affect relationships between manufacturing methods and variability, material properties, product geometry, and design requirement margins. In the design phase, material selection itself can impose consideration of material-specific failure modes that are naturally correlated to important attributes such as strength, weight, and geometry. ICME enables designers to quickly understand the complex and probabilistic interactions between the material, manufacturing processes, manufacturing variability, and design. Thirdly, it has been shown that successful account of variability of the manufacturing processes in life calculations leads to improved accuracy in declared low cycle fatigue crack initiation and damage tolerance lives on life limited gas turbine engine components. Furthermore, ICME enables engineers to rapidly explore more effective design and manufacturing solutions for delivering superior products at lower cost, faster but not without challenges. To highlight challenges and progress toward realization of this transformational technology, a survey of recent examples of materials and manufacturing process simulations along with the overarching approach and requirements within ICME to link these simulation capabilities to design and manufacturing methods will be reviewed from a gas turbine engine perspective.

Keywords: Integrated computational materials engineering Materials and process modeling Integrated design and make Design systems Gas turbine engine

REVIEW

Impetus - why we care

System benefits are lost when communication and technical data does not pass across functional lines. System failures occur when 'islands' or 'silos' are allowed to persist in product development. Fundamentally, the maturing discipline of Integrated Computational Materials Engineering (ICME) aims to integrate more than 50 years of computational materials and process modeling research into engineering systems to accelerate the design/make process through use of physics-based models and virtual linkage of the manufacturing processes with material structure and design performance. Though historical focus has been on development of digital/virtual toolsets, the success and practical implementation of ICME into product development systems requires and forces the cross-functional information flow and cultural change that is required to realize significant manufacturing benefit and improve competitiveness. Digitally integrating design and manufacture up front for iteration in analysis is much less costly than redesigning and replacement of products in service.Successful integration of ICME into product development systems promises order of magnitude reductions in product development time and cost through use of physics-based models which replace costly experimental iterations for optimizing manufacturing processes and component performance, as illustrated in Figure 1. The magnitude of benefit promised through systems level solutions, ICME, and the necessity for a step change in manufacturing competitiveness has primed industry for the next manufacturing revolution. In addition to the reduction in product development cost and timeline for engine development, life cycle cost benefits can be realized after entry into service.

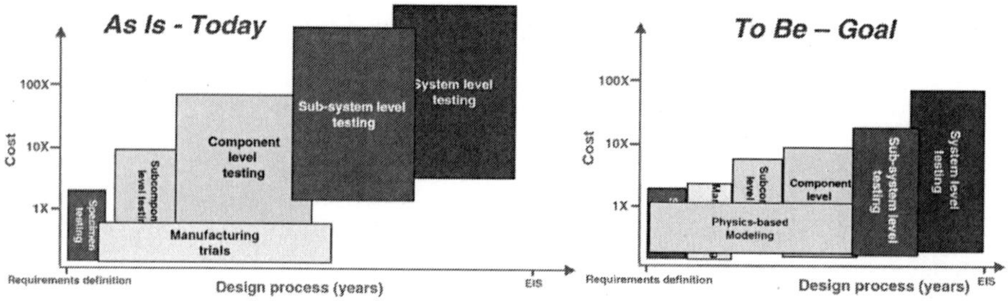

Figure 1 REDUCED ITERATIVE MANUFACTURING TRIALS THROUGH PHYSICS-BASED MODELING BENEFITS SYSTEM DESIGN.

Figure 2 illustrates the benefits of an improved approach to product development by introducing ICME earlier in the *design stage* after incorporating all relevant knowledge throughout the product life cycle. It can shorten the time to market and product capability responsiveness and deliver savings through reduction of recurring production and part costs. Improved accuracy in declared lives for life-limited components derived by the application of physics-based modeling of manufacturing processes can provide a significant reduction in cost of operation and maintenance of gas turbine engines. In addition, the application of ICME with Prognostic Health Monitoring (PHM) can also deliver significant business benefit by better informing maintenance decisions and overall fleet risk evaluations. Integration and application of computational materials engineering capability at Rolls-Royce has delivered significant business benefit in a number of ways including, but not limited to the following:

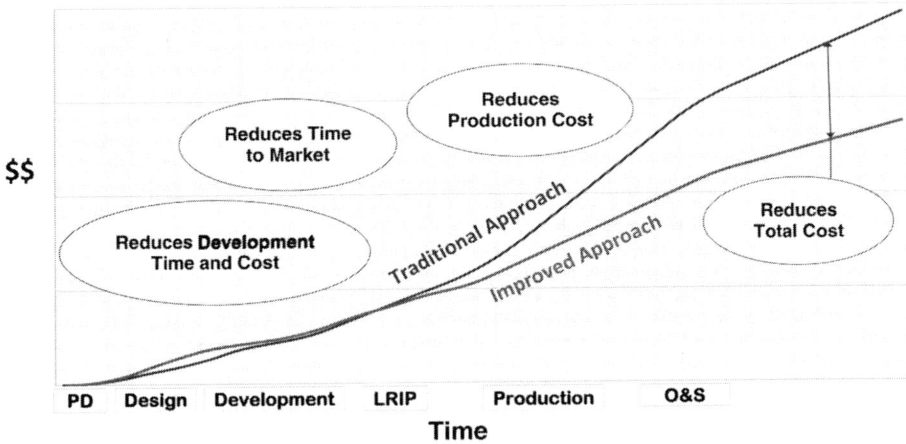

Figure 2 IMPACT OF ICME ON PART LIFE CYCLE COST.

- Distortion prediction of forged parts resulting in less scrap, fewer concessions (99.98% Right First Time), decreasing forging machining costs (50% cost reduction), and increasing stock-turn.
- Reduction in casting scrap (> > US$5 M p.a.).
- Increasing tensile strength (approximately 5% on forged/formed parts).
- Significant reduction (by 90%) of forming trials (powder HIP process) due to predictive part geometric accuracy.
- Optimization of parts in furnaces to increase utilization; increasing stock-turn and lowering costs.
- Shortening development time and time to market for new product introduction and redesign efforts.

The magnitude and opportunity of the benefit afforded by successful integration of ICME into product development systems is substantial. However, the implementation of ICME into product development systems has several significant challenges that must be addressed if bottom line business benefit is to be fully realized.

Scoping the ICME landscape

The integration of computational materials and process modeling into product development systems promises significant benefit for cost reduction, performance improvement, and lead time reduction across many material systems, commodity streams, and products, as well as across the product life cycle itself. As a result, a key challenge exists in appropriately focusing ICME efforts to match the greatest business benefit rather than trying to solve too broad a problem. Manufacturing companies do not sell technology; they sell products. Thus, all prioritization of ICME technology development and implementation efforts must have line-of-sight to bottom line business benefit and delivering superior product to market faster and at lower cost. ICME efforts can fail before they start simply because they have not been scoped for success and the focus falls solely on developing technology and not delivering to customer requirements and business benefit. Initially, internal design and manufacturing functions within the business can be engaged to articulate how ICME methods and toolsets can better deliver to their greatest product improvement needs. This facilitates prioritization of efforts to specific prioritized product benefits, delivered through specific components or sub-systems, and identifies specific design/make stakeholders that will become future business champions if the project is successful. For example, Figure 3 illustrates some general benefits/needs internal sub-system teams might articulate for a gas turbine [1].

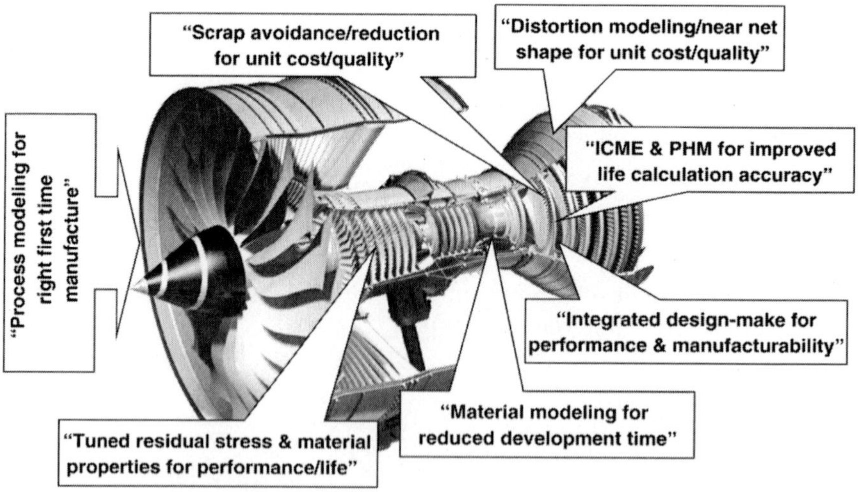

Figure 3. PRIORITIZATION OF EFFORTS BASED ON CUSTOMER ARTICULATED BENEFIT.

In addition to continual customer engagement, it is important to consider several other scoping factors for ICME effort prioritization. The selected effort must be planned to deliver a phased benefit with a series of quick wins early on to maintain business support. This may mean that the selected project may be the one with the greatest business impact, or the project that delivers the largest benefit for an acceptable level of risk. Furthermore, overlap in benefits or needs articulated from customers may highlight efforts that would have a broader impact in areas across the business. Wherever possible, planning to develop once and reuse multiple times will pay big dividends when scaling up the benefit and may highlight a lower risk project that can be selected as a pilot while providing the foundation for future benefit on higher value components and systems. For example, engagement with design, component structural analysis and lifing, and materials and manufacturing functions internally at Rolls-Royce led to grouping ICME's capability into commodity buckets with similar manufacturing process applicability as highlighted in the tabs of Figures 4 and 5[2].

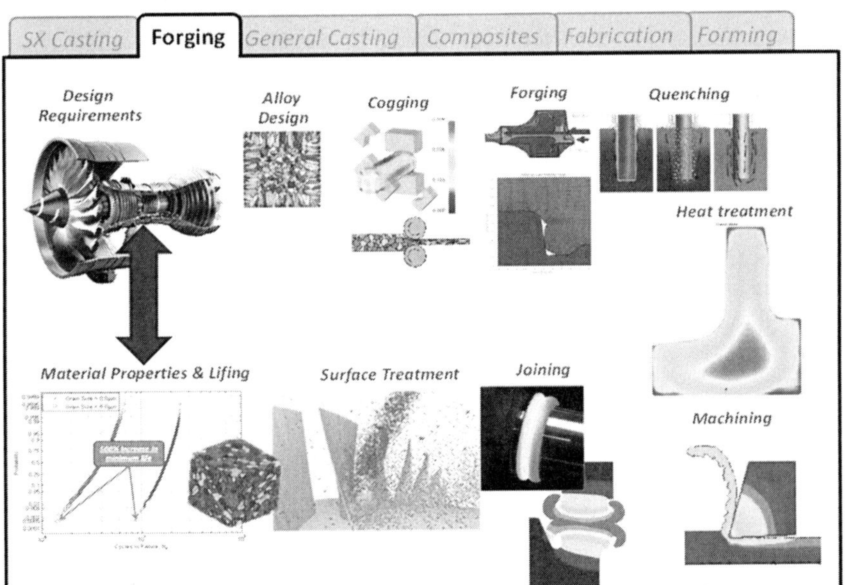

Figure 4. EXAMPLE COMMODITY MODELING MAPS OF A FORGED PART.

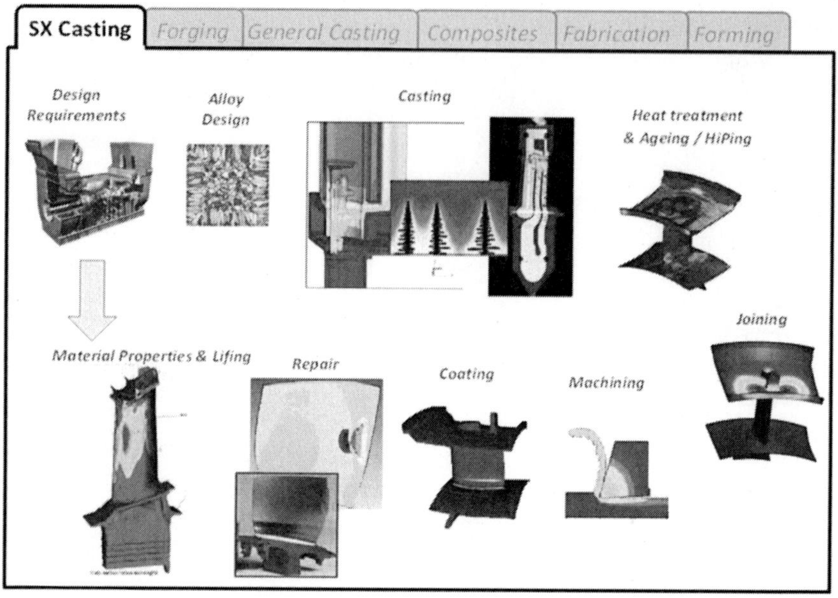

Figure 5. EXAMPLE COMMODITY MODELING MAPS OF A CAST PART.

These process modeling maps provide a graphical illustration implying the sequence of ICME models that virtually link design requirements, the manufacturing processes that make up a product's 'digital pedigree,' and the predicted material structures and properties that can predict the component manufacturing yield and design performance. This gives context for key stakeholders to quickly contextualize the larger ICME landscape for a given commodity group and highlights a digital thread for each component that links back to the design requirement articulated from customer engagement.

It is important to note that the integration of ICME into product development systems is not just digital but physical, heuristic, and relational as well. As a gas turbine engine OEM, Rolls-Royce delivers world-class power system and service solutions to our customers... but not on our own. This requires a trusted ICME supply chain and infrastructure of incentivized technology partners that can collaborate to deliver what is best for Rolls-Royce, as an original equipment manufacturer (OEM), and for the product development team as a whole. The interdisciplinary nature of systems engineering underlines the criticality of selecting an appropriate cross-functional team (i.e., design, materials, manufacturing, supply chain, and export control/IP) for program definition up front to ensure that programs are planned to deliver the right information to the right product development team members at the right time.

Though ICME concepts are not new to the aerospace industry [3] and academic work to develop isolated computational material models has been going strong for several decades [4], the competency of the existing supply chain in integration and use of ICME methods and capability is still relatively immature. To facilitate more rapid adoption of this technology across the industry, government and industry must fund, sponsor, and partner in activities aligned to greatest business benefits. To this end, Rolls-Royce continues to invest in developing and deploying new technology through strategic partnerships with a global University Technology Center network, Advanced Manufacturing Research Centers, and public-private partnerships like the Metals Affordability Initiative [5]. An illustration of the strategic partners and development and deployment activities for a healthy ICME supply chain are highlighted in Figure 6.

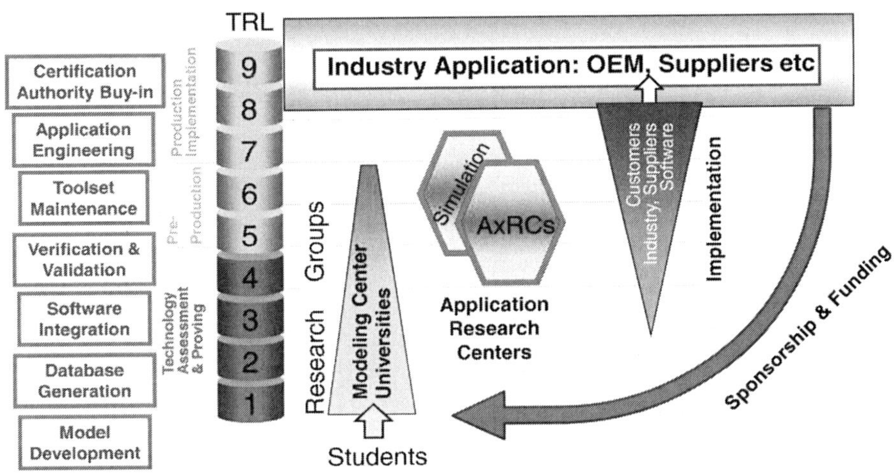

Figure 6. ICME SUPPLY CHAIN FOR DEVELOPING, DEPLOYING CAPABILITY FOR INDUSTRY CONSUMPTION**[1]**.

Systems integration

To further scope the ICME opportunity and challenge, it is important to understand that ICME is not and should not be restricted to only design. For example, Rolls-Royce has a vision to embed ICME across the entire product development, introduction, and life cycle management process, including the following:

- Preliminary design (more analysis at sub-assembly and engine level to eliminate risk early in the design process).
- Detailed design (use of analysis-based optimization and robust design and manufacture to enable rapid definition at component and sub-system level).
- System design using high-fidelity virtual engine (more sophisticated, multiphysics analysis to accurately predict engine behavior linking manufacturing variability to component-specific performance).
- Virtual manufacture (optimization of those aspects of manufacturing that have no impact on the product design but may affect, for example, the cost).
- Virtual testing (analysis-based test strategy, planning, and correlation to reduce the need for repeat testing).
- Virtual product validation/certification (rapid certification based on validated analysis, simulation and modeling).
- Product life cycle analysis (fixed design, probabilistic analysis, and updating of models used during development for improved service and aftermarket decisions involving business risk, such as maintenance and product improvement costs).
- Process capability analysis (data and knowledge capture for reuse in continuously validated and improved methods, constraints, and rules used above).These describe the elements of larger virtual product development system for faster delivery of superior products to market at lower cost. Figure 7 illustrates a 'six-stage' process defining the life cycle of a product and how each of these elements roughly maps onto it.As illustrated by the dashed lines in Figure 7, both legacy and new product development efforts generate significant data and knowledge that can be leveraged in unique and innovative ways at different points in the life cycle to deliver benefit. However, with the addition of more complex analysis, the accompanying large volume of digital data that will be generated out of such complex systems and the pragmatic reality that uncertainty will pervade data and model inputs, the ability to manage risk and effectively utilize these new capabilities to inform engineering decisions requires an appropriate framework and mindset.

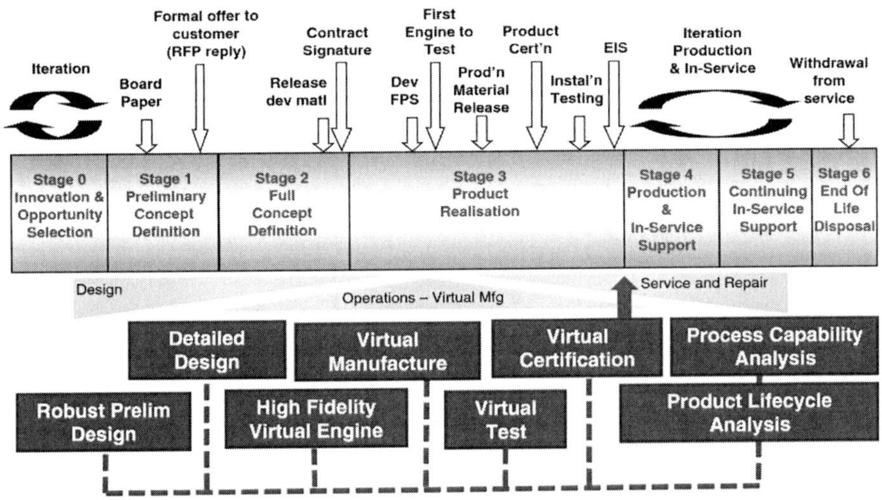

Figure 7. SIX-STAGE PRODUCT LIFE CYCLE AND VIRTUAL PRODUCT DEVELOPMENT ELEMENTS[1].

Facilitating engineering decision making: informing decisions under conditions of complexity and uncertainty

Managing the development of complex new systems requires many decisions using combinations of analytical engineering models, experimental data, and expert knowledge. This process is inherently expensive when risks are high and is made even more expensive when risks are not effectively managed throughout the process. Though new systems may ultimately be successful in delivering product to market, the path to success is often far from ideal in terms of the use of models, data, and knowledge to manage risk. Models, data, and knowledge are all valuable resources in their own way but are often pitted against one another inappropriately at key decision points or worse, trusted implicitly or ignored altogether. Decisions must be made, for example, about what to analyze, what to test, what conditions to test, whether a new technology should be inserted, and the level of model fidelity needed.

Significant reduction in the development time and cost depends on quantifying and responding appropriately to risk as early during the process as possible. An error in material selection or manufacturing process definition caught early during design is not nearly as costly as the same error caught later. There are, of course, several reasons such an error might not be caught early. First, the breadth in ramifications may simply not be taken into account. For example, the material may be selected based on a requirement to operate at higher temperatures but other requirements such as corrosion resistance, manufacturability, and inspectability may not be adequately considered or known. Secondly, an understanding of the typical behavior of the material or manufacturing process may be based on relatively small samples or idealized conditions. Later when the true variability becomes apparent, it becomes much more difficult to manage. In both examples, decisions are made under conditions that are unnecessary given many of the tools available today. Physics-based models such as those envisioned under ICME are being used to quantify not only typical behavior but also variability. In addition, today's access to affordable computational resources is making uncertainty quantification (UQ) for large complex systems possible when addressing the breadth issue.

However, quantification and management of risks associated with complex new systems mean making UQ and decision-making tools and training available to the right people throughout the process. Uncertainty and therefore risk tends to be highest at the onset of development and will be continually managed over time. What is not as well appreciated is the impact of sub-optimal paths on cost. The path used to blend and continually update engineering models, experimental data, and expert knowledge into information useful for decision making is critical to reducing development time and cost.

Finally, the same arguments made for reducing development costs may be advanced for reducing life cycle costs. Despite the enormous investments made to reduce risk and uncertainty prior to production release, a great deal is still learned over the life cycle of the product. New data is gathered on product performance, operational usage, changing economic conditions, for example, which continually present challenges to the decision-making process. Win-win situations can be made for OEMs, suppliers, and customers alike simply through the effective use of updating methods (e.g., Bayesian or otherwise) and decision-making tools.

ICME in context

Successful integration of ICME into next-generation product development systems will involve distributed (i.e., cross-supply chain) problem-solving solutions the ICME supply chain can *understand*, *trust*, and *use* efficiently and effectively to deliver a superior product to market faster and at lower cost.

Understand it

ICME system elements must be developed and delivered in a context and framework that product development teams can understand. Figure 8 highlights some of the key system elements involved in a product development system.During the design process, a cross-functional integrated product development team will exercise these system elements to develop and deliver a product solution. To contextualize how capability is delivered into a digitally integrated design and make system, consider the high-level engineering workflow at the sub-system level of a forged component highlighted in Figure 9. Figure 9 emphasizes the cross-functional and cross-supply chain interfaces that must be managed efficiently and effectively to facilitate truly concurrent engineering earlier in the product development process.

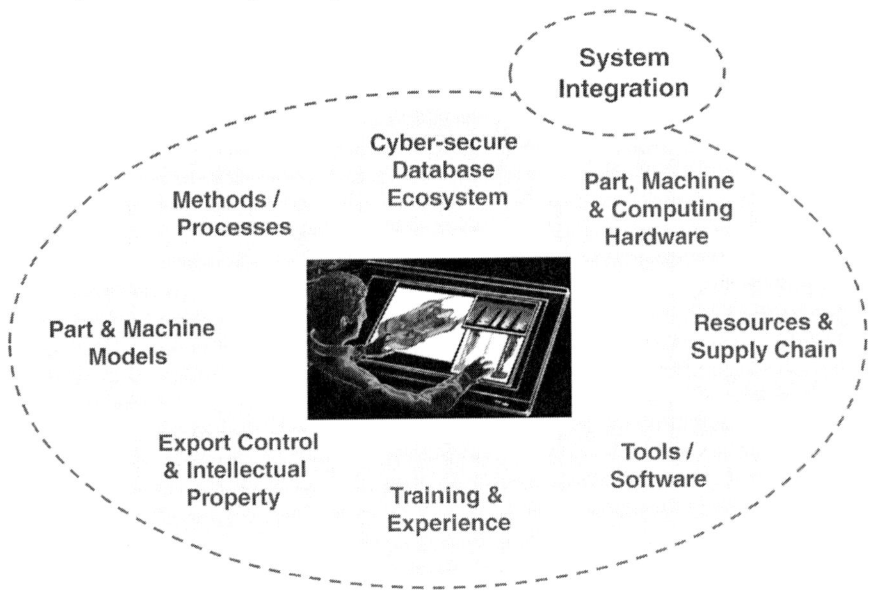

Figure 8. FOUNDATIONAL ELEMENTS OF A PRODUCT DEVELOPMENT SYSTEM.

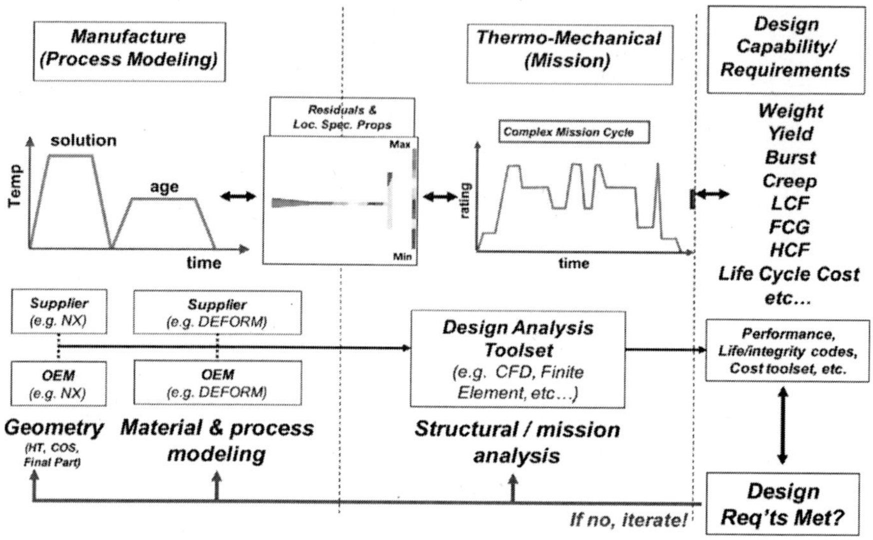

Figure 9. HIGH-LEVEL ENGINEERING WORKFLOW FOR FORGED PART DESIGN**[1]**.

Computational material models and process models deliver predicted component performance in terms of residual stress and location-specific material properties that now enable designers to better understand the impact of manufacturing processes and manufacturing capability on the resulting design space. Digital integration and automation of the engineering workflow enables rapid exploration of feasible design space (based on known material and manufacturing capability) while simultaneously identifying robust methods of manufacture to deliver 'Right First Time' solutions that meet the requirements while minimizing or eliminating expensive physical trials. The integration of new ICME capability into workflows for components will require additional work by the engineering function in an environment that already has extreme time pressure to deliver results. Thus, ICME development efforts must address a number of design requirements and needs, or time pressure and frustration will prevent use and realization of benefits associated with the additional capability. To this end, ICME models, methods, and software toolsets must be as follows:

- *Accurate enough* Inaccuracy can result in unanticipated life cycle costs, but too much attention to accuracy results in overly high development costs. Both ultimately represent risk that needs to be quantified and managed. This is one area in which uncertainty quantification, ICME, and decision analysis tools become particularly useful. Early in the product life cycle when understanding system, sub-system, and component level requirements is especially challenging, such tools enable key decision makers to quantify what is 'accurate enough' to move forward in the development process. Accuracy must be considered both qualitatively and quantitatively. The qualitative component of accuracy (the pattern of behavior) must be consistent with physical reality to appropriately capture trending. Quantitative accuracy depends on several things including, but not limited to, the modeling algorithm, input data accuracy, material/process variations, and measurement accuracy for validation efforts. In some cases, in-process sensing and control can be used to compensate/mitigate quantitative inaccuracy. Model development efforts should be limited to the accuracy needed for the specific product or application of interest. A clear understanding of the requirements up front will define what accuracy is 'fit for purpose' and scope development efforts to deliver an appropriately refined engineering solution.

- *Efficient and usable* Engineers are fighting a battle against timescales for program delivery, and the inclusion of new capability is asking them to do more with no additional calendar time. Therefore, new capability must be delivered for seamless integration into existing design systems with as little additional computational and/or calendar time as possible.
- *Relevant* Though most material models must operate at smaller length scales (micro- or nanoscale) to capture appropriate mechanisms, delivered solutions must be compatible with continuum scale which is where most engineering analysis occurs.
- *Available* Export control and intellectual property implications with sharing models and data must be reviewed and planned for up front. This ensures that the developed solution has all required business agreements and export licenses in place for the cross-supply chain product development team while fully complying with export regulations and legal arrangements. Furthermore, a cyber-secure information technology collaboration platform/solution must be available to streamline access to models and data while ensuring export and intellectual property right compliance.
- *Validated* Technical and business benefit validation must be considered. Manufacturing variability and uncertainty must be addressed to build confidence and trust in predicted results; however, validation of the ICME capability to deliver bottom line business benefit is what will transition from technology push to business pull.

- *Maintainable* Developed solutions must be maintained, version-controlled and updated as improvements are made or system upgrades occur. A competent and robust ICME supply chain is required to support maintenance of developed solutions or they will cease to be useful.
- *Interoperable* There are many different product development solutions and software used across the supply chain currently. This poses a challenge for streamlining digital communication between each organization. To this end, ICME efforts must deliver generic, transferrable, and modular solutions that are cross-supply chain compatible and not OEM or supplier specific.

Trust it - verification and validation

Manufacturing variability and uncertainty throughout product development is real and inevitable. Not everything is controlled and/or known which leads to substantial risk if not acknowledged and managed accordingly. In addition to verifying developed solutions to give expected outputs, validation that predicted results line up with reality requires an account of variability and uncertainty inherent to the manufacturing process.In line with best practice robust design and make principles, it is important to identify, prioritize, and quantify sources of variation and uncertainty that impact key process variables (KPVs) in manufacturing, material structure, and ultimately product characteristics and performance in design. Conventional robust design toolsets (e.g., fishbone diagrams, p-diagrams) can be used to help engineering teams identify sources of variability and uncertainty. Expert opinion, available prior knowledge/data, and statistical models and tools can then be used to prioritize key drivers for performance and cost (i.e., Pareto chart, QFD). As is often the case when uncertainty is high, expert opinion needs to be combined carefully with analysis models and available data. This can be done working within a Bayesian statistical framework which then provides a general foundation for incorporating new data, driving toward reduced uncertainty along with verification and validation. The use of ICME toolsets and strategic testing can be leveraged to formally quantify, understand, and reduce variability and uncertainty to improve results to desired customer goal with an acceptable level of confidence. Figure 10 graphically illustrates the general process steps of manufacturing variability and uncertainty quantification and reduction.

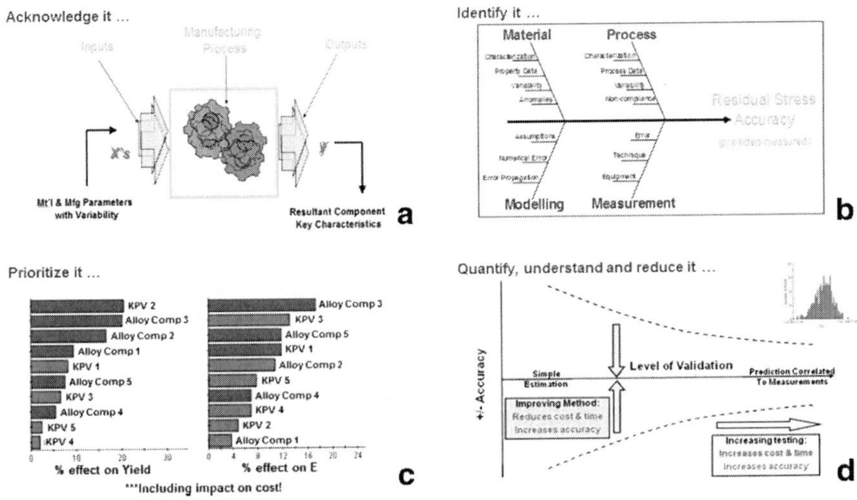

Figure 10. GENERAL PROCESS STEPS OF MANUFACTURING VARIABILITY AND UNCERTAINTY QUANTIFICATION AND REDUCTION. Illustration of process to **(A)** acknowledge, **(B)** identify, **(C)** prioritize, and **(D)** quantify manufacturing variability and uncertainty in support of robust design and make processes [1].

Comprehensive implementation of ICME, especially at the engine system level, will require systematic, rigorous, and quantitative verification and validation (V & V) efforts, including targeted demonstrations. Though beyond the scope of this paper, a comprehensive overview of V & V challenges and progress as related to ICME has been captured in a paper by Cowles, Backman, and Dutton [6]. To establish confidence in application of ICME at any system level, an appropriate V & V plan must be established and executed to ensure that the modeling methods have been vetted to the level of accuracy required for the target application.

Use it

To provide specific context for integration of ICME into product development systems, a survey of the opportunities and challenges for use of ICME at different stages in the life cycle will now be reviewed.

Lifecycle integration examples and challenges

Preliminary design

Preliminary design occurs during phases 0 and 1 of the six-stage product life cycle (Figure 7). Stage 0 can occur over an extended period of time as the product requirements and the product attributes are developed simultaneously. In stage 0, product attributes of interest for gas turbine engines are performance, weight, cost, and life. When working closely with a customer, the two questions 'What can you do for me?' and 'What do you need from me?' are answered.

It is in this stage that the required material properties for a product are identified. ICME becomes integral to the product during this stage of the product life cycle. For example, the differing material properties for the rim of a turbine disc compared to the bore of the same disc are identified. The mathematical identification of these properties enables a robust design solution for product design.

Mathematical methods using design of experiments explore more solutions more thoroughly. The range of solutions establishes a design space where a product design can be selected that is less sensitive to variations. The design solution evolves from a top level solution where the design is established as an overall architecture, to detailed designs of sub-systems and components. This stair-step approach makes for not only effective product design by exploring the design space but also efficient solutions by keeping the solution from going off track. To include detail modeling of each component at the architectural selection phase would create a model too large to run on most computers and take too much time to set up such a model. The human interaction with the models between the levels of optimization needs to be effective in keeping the solution process on track toward a truly optimal product design.The design of a component is a combination of the geometrical features and the material capability. Prior to ICME, the design of material capability was limited to selection of a material from a database with experimentally measured and validated nominal properties. ICME dramatically improves the design of a component by permitting the simultaneous detail design of geometrical and material features (microstructure and properties). By adding material features to the design space, the robustness of the product design is improved with the intent of reducing the chance of in-service issues. Figure 11 illustrates the cost and time benefits of getting a design right at the analysis phase.

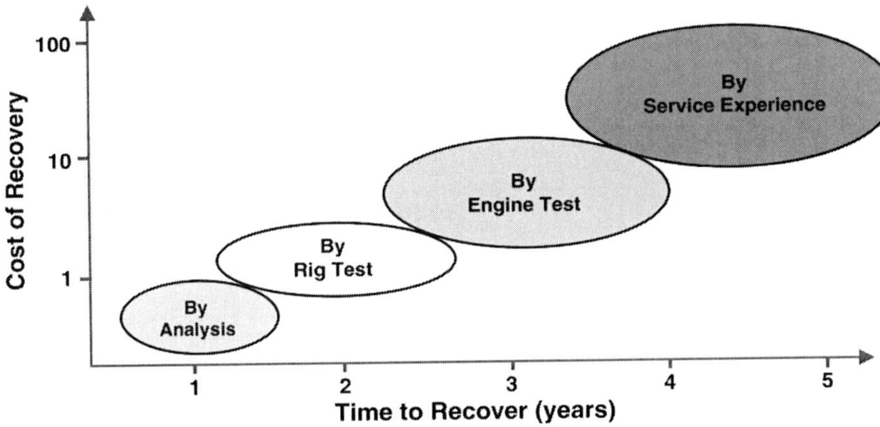

Figure 11. BENEFITS FOR OPTIMIZING THROUGH DESIGN BY ANALYSIS EARLIER.

Robust design requires mathematical representation of material properties as well as geometrical features and applied loads. ICME provides not only material capability in mathematical terms but also delivers location-specific properties that can be used to design and produce a product with superior attributes.Rolls-Royce has been effective in performing robust design using Isight™ as the integrating framework for a number of programs as illustrated in Figure 12. From the first performance models, Rolls-Royce includes requirements for modules for engine modules and components as well as aircraft-required attributes. The outcome is an optimized, robust design. Numerous solutions are studied with varying requirements and defined optimizations to understand the design space. As the customer and Rolls-Royce gain confidence in the product strategy, the design iterations become more focused on the details of the modules then the details of the design. ICME is integrated with the design approach; first by various validated design allowables early in the process then by detailed process models. The result is an optimized product design.

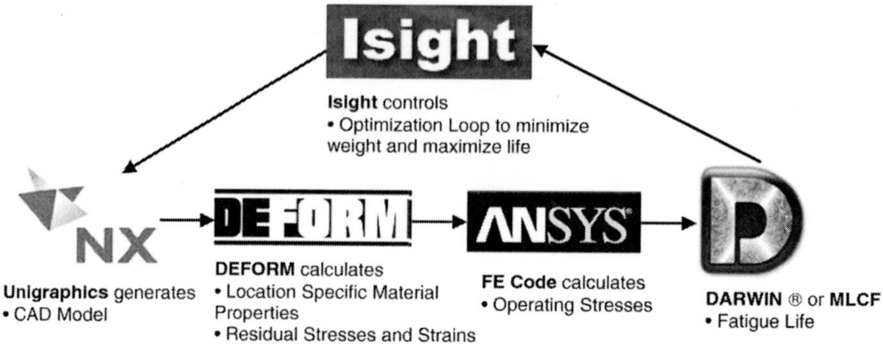

Figure 12. ISIGHT ™ DEFINE AND CONTROL THE INTERACTION OF NUMEROUS DESIGN PROGRAMS[1, 2].

Materials design

Feasibility evaluation, development, and insertion of new and optimized materials technologies have become heavily dependent on the use of ICME tools and methods, as experimentation and empirical approaches alone cannot keep pace with current advanced aerospace component design cycles. The development and growth of computational tools for materials engineering have historically been hampered by the complexity and diversity of phenomena and properties that must be captured in order to deliver a robust and integrated engineering solution. Fortunately, ICME tools and the computing power required to process complex and simultaneous events during manufacturing and in service are now reaching a level of maturity where they can have a substantial impact if they can be integrated into product development [7].

A common and successful application for ICME tools has been the prediction of multicomponent alloy solidification behavior and phase equilibria in order to more efficiently assess novel compositions for structural materials. For example, innovations in existing nickel-based superalloy compositions and processing routes are required in order to overcome the challenges of usage at higher temperatures in advanced and future aerospace gas turbine engines. The most significant materials limitations for advanced γ/γ' (γ'') alloys in particular are phase instability above 700°C and mechanical strength reduction from segregation-induced defects. Eutectic γ/γ'-δ alloys (Figure 13) may offer a novel solution due to the additional reinforcing phase δ, which has demonstrated stability in experimental studies up to 1,000°C [8].

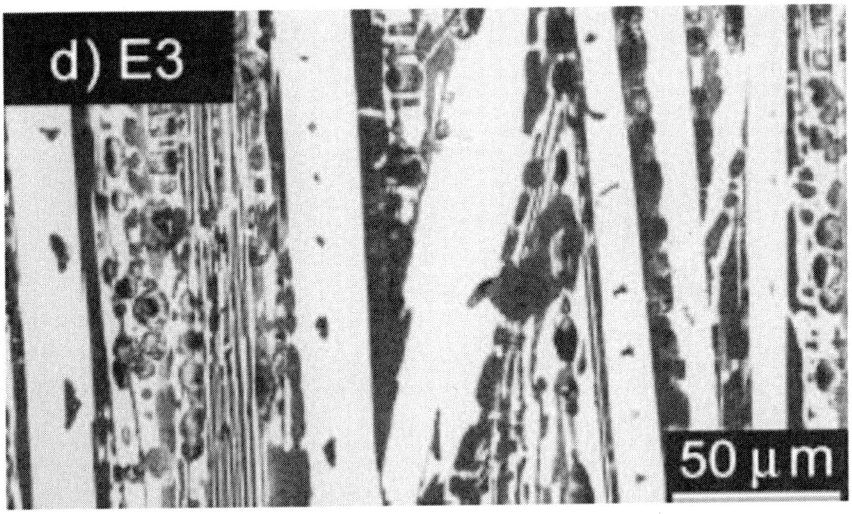

Figure 13. REPRESENTATIVE MICROGRAPH OF A Γ/Γ′-Δ EUTECTIC ALLOY, SHOWING Δ-PHASE PLATES WITHIN A Γ/Γ′ MATRIX[8].

The rapid assessment of these potential next-generation alloy concepts has been made possible by the use of ICME tools. A commercially available software, with the incorporated Ni-alloy thermodynamic database, has demonstrated the ability to predict and confirm trends in primary phase volume fraction (Figure 14) and to confirm qualitative trends for phase equilibria and solidification behavior resulting from composition changes in multicomponent γ/γ'-δ alloys (Figure 15) [9]. Comparison to experimentally derived results has shown good agreement in some instances, whereas areas of disagreement between the experimental and predicted metallurgical data highlight gaps in the current thermodynamic database that require further optimization. Even given these current limitations for predictive capability, these particular ICME tools have resulted to a better understanding of the reasonable design space for a ternary γ/γ'-δ alloy system and provides the foundation for future exploration of these novel structural materials.

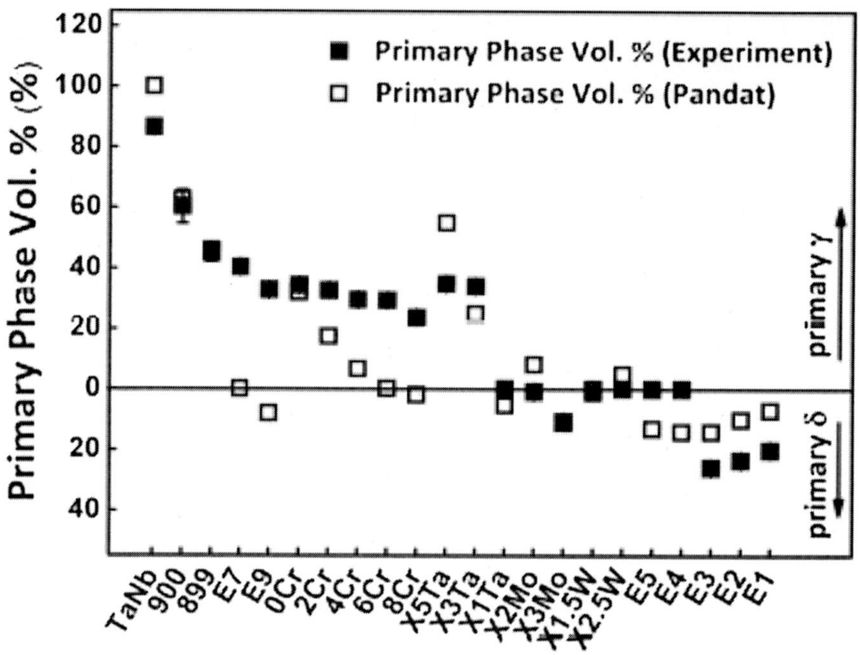

Figure 14. EXPERIMENTALLY DERIVED AND CALCULATED PRIMARY PHASE VOLUME FRACTION IN VARIOUS ALLOYS. The trend to change from primary γ to δ with increasing Nb content is illustrated [8].

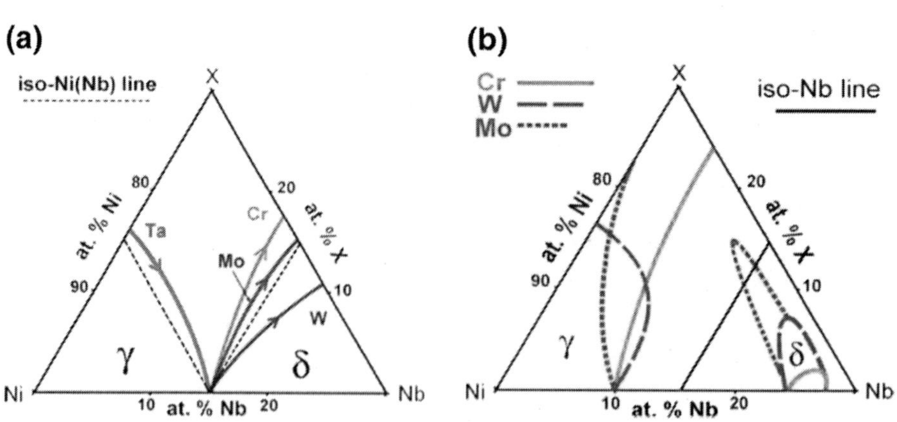

Figure 15. SUPERIMPOSED LIQUIDUS (A) AND ISOTHERMAL (B) PROJECTIONS OF NI-NB-X TERNARY PHASE DIAGRAMS. The different trends in the γ to δ eutectic trough and solubility in various alloy systems are illustrated respectively [9].

Detailed design

In today's highly competitive gas turbine industry, robust multidisciplinary design tools are required early in the design process which links several analyses to achieve aggressive overall system goals. An example of this detailed design process for a critical rotating component is shown in Figure 16. ICME is a relatively new addition to this design iteration loop and is highlighted in orange. A robust tool for sizing a new rotor disc would traditionally link aerodynamic, mechanical, and fluid systems design trades to the resultant heat transfer analysis, structural analysis, and predicted service life of the component. In this manner, design trades are quickly assessed, and the impact to the sub-system is quantified. Linking steps in the analysis allows perturbation of the system by making, for example, a change to a secondary fluid cooling scheme. Such a change would alter the predicted disc temperatures, resultant stress, and life prediction as well as change in blade tip clearance. Linking these types of tools can provide the design team an updated performance, life, and cost prediction indication in minutes.

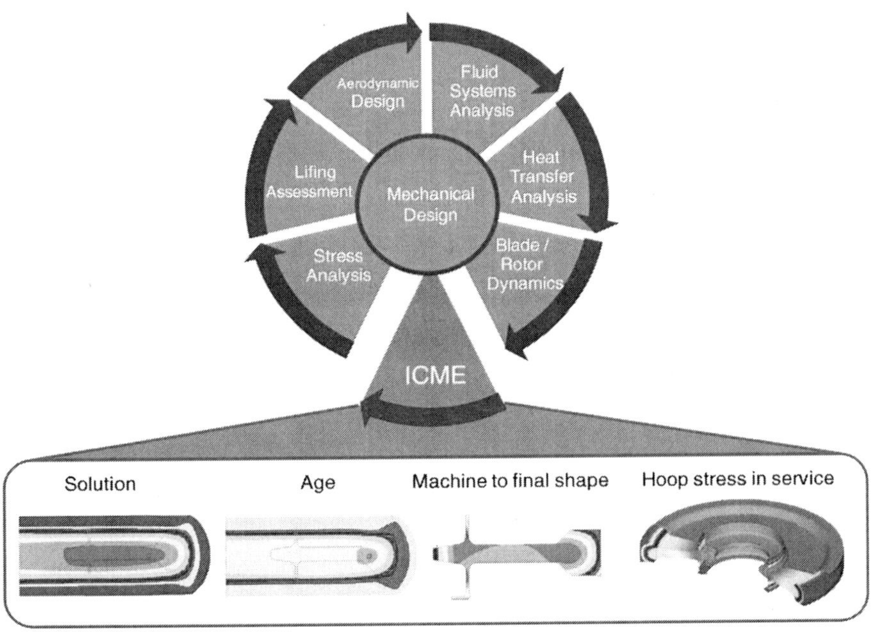

Figure 16. GENERIC EXAMPLE OF DETAILED DESIGN PROCESS FOR A CRITICAL ROTOR COMPRESSOR BLISK[10].

With the advent of high-performance computing, linking these design tools becomes more attractive and affordable. Recent advances in materials modeling have resulted in analytical tools that package nicely into the design loop described previously. Combining with the traditional design loop, high-performance computing with quantitative material property and residual stress modeling becomes a powerful toolset for advancing the state of the art. Figure 16 shows the ICME modeling capability which are now included in the detailed design process. This approach was recently applied to a legacy LCF spin rig disc experiment. In this historical spin rig study, different initial residual stress profiles were imparted to discs by applying varying degrees of a single pre-spin event prior to cyclic testing. These pre-spin events created regions of local yielding, leading to localized compressive residual stress. Components were cyclically tested to crack initiation and fracture with and without beneficial residual stresses imparted due to pre-spin at critical locations. A 2.8× increase in total life to failure (crack initiation and crack growth) was observed for the pre-spun components compared to the baseline condition. Further details of this experiment are provided in depth by Shen et al. [10]. Application of ICME toolsets was used to predict the component residual stresses by modeling the process of forging, machining, and pre-spinning operations performed prior to testing. Using inelastic stress calculations, the improved total life to failure observed in the spin test study could be predicted through the Rolls-Royce analytical lifing methodologies. Based upon the validation testing conducted previously on Rolls-Royce historical rigs, it has been shown that the accuracy of total life calculations can be improved by accurate account of induced residual stresses. By applying ICME toolsets, the residual stresses that form during the forging and heat treatment process can also be derived for a more accurate description and analysis of the finished component. These calculated residual stresses can then be combined with thermo-mechanical mission stresses for analytical low-cycle fatigue initiation and fatigue crack growth life calculations. Successful control of the manufacturing process allows for tailoring of the final component residual stress and potential for total life improvement in low-cycle fatigue crack initiation and fatigue crack growth.

The work done by Shen et al. [10] opens up the possibility of tailoring the forging shape and/or heat treatment cooling rates to achieve an optimized component solution. Such solution may be minimum weight - a full life component design that meets the sub-system goals. Figure 17 shows the disc used in the previously described spin test with two sample forging shapes. This figure illustrates the variables that can be controlled to tailor the material properties and residual stress of the finished part. Namely, the local heat transfer coefficient (as a function of time) and forging geometry are the key process variables contributing to the resultant predicted service life. This becomes an ideal problem for DOE and/or optimization techniques. Rolls-Royce has recently conducted this type of work for several critical components in recent advanced system demonstrators to maximize service life. This required close working with suppliers to achieve an overall acceptable solution for the program. The result was a specific forging geometry and heat treat fixture that provided acceptable material properties and desired initial residual stress profile. In addition, predicted deflections, obtained by simulating the machining process, were very close to what were observed in manufacturing. Predicted deflections from this modeling were used to guide manufacturing to deliver a complicated component to print without deviation.

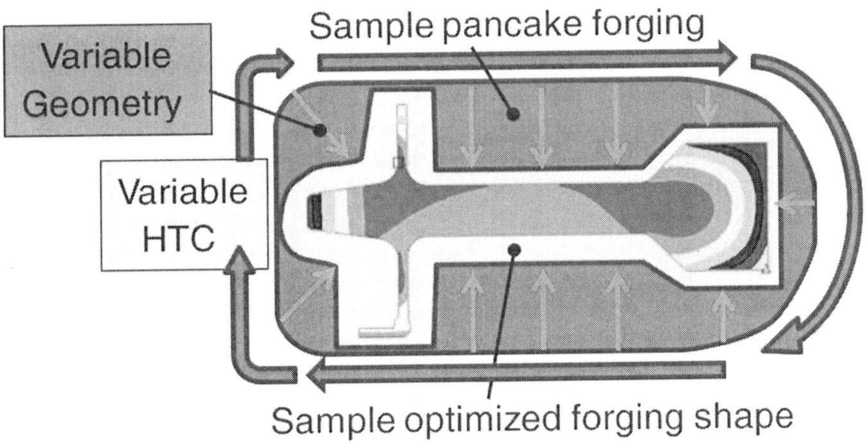

Figure 17. SCHEMATIC HIGHLIGHTING PROCESS VARIABLES DRIVING MATERIAL PROPERTIES AND RESIDUAL STRESS.

The introduction of ICME techniques to the design of gas turbine engines presents a potentially game-changing technique for designing critical engine components for higher life. By considering the residual stress inherent to any forged and heat-treated component allows the designer to make more efficient use of material, weight, life extension, and/or cost design trades. For existing products, this approach can be applied to extend the life of fielded components through alterations to the heat treatment process or forging geometry only, limiting the cost impact to implement changes. Such changes are often warranted later in the design life cycle as engineers gain confidence in how components are actually used and maintained in service. In addition, understanding of the inherent manufacturing process variability can produce a more robust assessment of fielded component risk. This increased understanding can be used by the fleet operators to improve decision making on maintenance and overhaul scheduling in addition to reliability predictions. The benefits of realizing and incorporating this technology are just now beginning to come to fruition. Rolls-Royce is now updating their design methodology to include material and process modeling in the design process. While much work remains to fully benefit from the potentials of including ICME into our design practices, initial work has shown positive trends in capturing additional life and minimizing part weight.

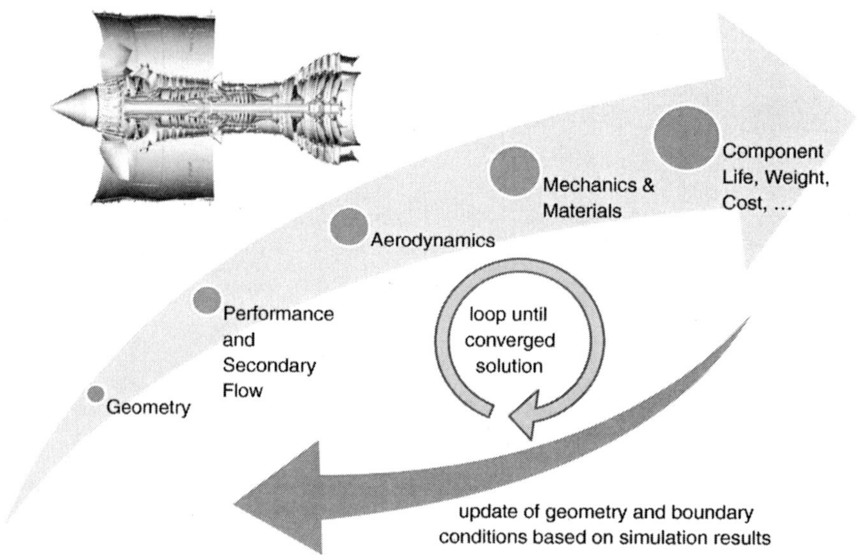

Figure 18. SYSTEMS LEVEL VIRTUAL ENGINE MODELING FOR ASSESSING IMPACT OF SYSTEM VARIATION.

Virtual engine

Detailed design and analysis of individual components and sub-systems alone are not sufficient to fully realize/maximize aero engine performance. Improved design understanding of the entire system is required in order to optimize the final product according to environmental and operational constraints. Significant recent advances in high-performance computational facilities enable analysis of many sub-systems in one 'product level' model with sufficient geometric detail and boundary condition definition required to simulate system behavior, such as component movements under steady-

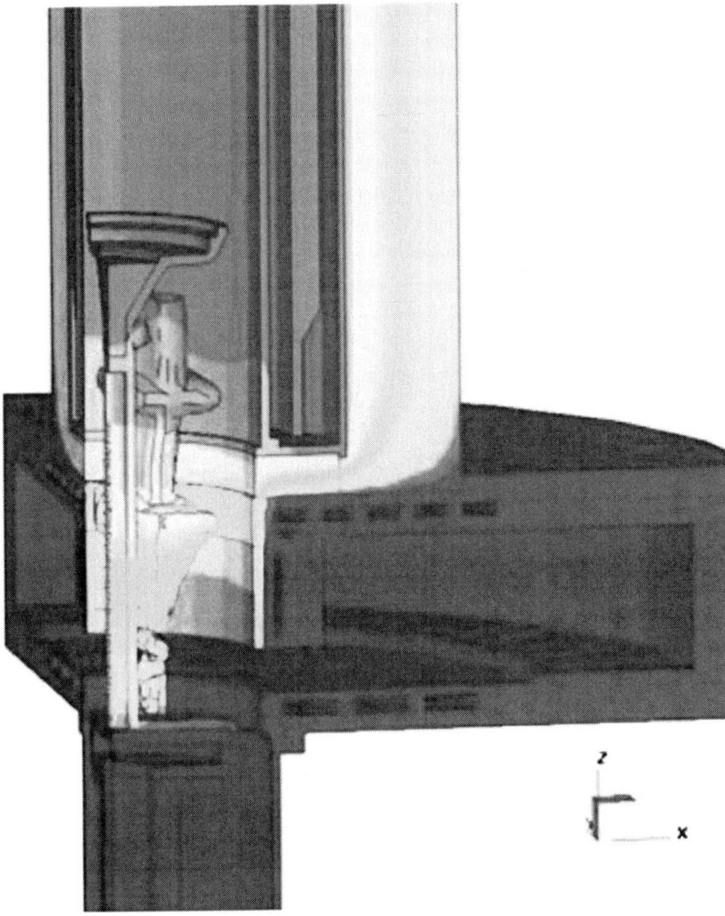

Figure 19. CASTING MODELING FOR PREDICTION OF YIELD FROM MANUFACTURING PROCESS. Shows temperature contours on a 90° segment of a partially withdrawn casting assembly within a Bridgman-type furnace.

state and transient loads. Many of today's design tools do not sufficiently account for material and manufacturing variability and their impact on component and assembly shape, structural integrity, and performance. For gas turbine design, system level 'virtual engine' models incorporating multidisciplinary analysis (i.e., aerodynamics, mechanics, materials, and performance) are giving greater insight to the impact of system variations on overall system performance, cost, weight, and life (Figure 18) [11].

Virtual manufacturing

Manufacturing is traditionally treated as an entirely separate function within the life cycle of a product. *Design engineering* will perform studies and develop various iterations of a product design then deliver a final definition to *manufacturing*. This definition may be in drawing or electronic formats. *Manufacturing engineers* will then determine a method to manufacture the product. In case of a new product design for a machined component, a manufacturing engineer will define a sequence of operations utilizing machine tools, fixturing, cutting tools, and inspection equipment based on available experience and knowledge. Once a method is defined, tooling/fixturing has been designed and built, and NC programs have been written, the process of prove out begins. Each step in the method will be proven out on the production machine tool. If any step in the method does not perform as planned, then modifications are made to the process and tried again. This can become a lengthy and costly loop, especially when there are numerous machining operations. Inherent variations within the process can make troubleshooting of issues very difficult and time consuming. During these prove-out steps, production machine tools are being utilized for development and not for daily production. This reduces operational productivity, disrupts material flow, complicates scheduling and could lead to late deliveries. Furthermore, delay of validation testing of components (required to be conformed to final production processes) may also result in certification delays and additional costs.

The integration of technologies like ICME and model-based definition (MBD) into the product development process facilitates digital integration of manufacturing knowledge, standards, methods, capabilities, and limitations into the design development process. This allows engineering 'design and make' teams to predict, optimize, and prove out a components method of manufacture before any step of the manufacturing process is started. The condition of supply definitions can be optimized to minimize variations within a chosen method of machining sequences. Fixturing can be designed to provide the best holding capability and minimized part distortion. NC programs and cutting strategies can be fine-tuned for performance, capability, and robustness. All of these steps can go through numerous iterations virtually to achieve the most optimized process that will perform as predicted, and much of it can be done even before the design is finalized. When fully mature, this approach allows for a new component to be introduced into a production line Right First Time with a process performing at, or better than, the capability of existing processes.

To fully realize this vision, however, integrated tools that span the full supply chain are needed to achieve full capability. Robust modeling tools should be integrated with the Computer Aided Manufacturing systems to allow for full material characteristics prediction and optimization. Challenges arise when an external supply chain is involved, however, since systems that are internal to a company are much more straightforwardly integrated. The overall supply chain needs the ability to integrate their systems and processes to enable true lifecycle integration. IT security, export control and compatibility issues all add to the complexity of a global 'digital' supply chain.

Virtual simulation of the machining process exists today in most of the modern CAD/CAM systems and external applications such as Vericut™ by CGTech; however, most current manufacturing process simulations do not integrate available ICME toolsets for prediction of manufacturing process impact on part yield, material properties and/or design performance. Vericut™ for example can simulate the complete kinematics motion of a multiaxis machining center and interaction of the cutting tools and fixtures. Vericut™ can also simulate material removal by cutting tools and compare the cutter path resultant to the original part model. These simulations are valuable for a virtual prove out of the cutter path accuracy and potential interferences between tools/fixtures/machines/parts; however, these simulations are static and will not take into account cutter deflections or part movement from relieving residual stresses as material is cut away. To realize the full potential of an integrated virtual manufacturing system, ability to predict and simulate how the materials interact and react during an operation or process is critical. Fully integrated virtual manufacturing systems could eliminate the costly physical prove-out stages in manufacturing and facilitate Right First Time solutions when integrating known process capability to design-for-manufacture toolsets earlier in the product development process. New products could be introduced into production along with existing products, allowing manufacturing to improve management inventory while maintaining the 'heartbeat' of the product flow through the physical manufacturing facility itself.

As integration of virtual manufacturing across the supply chain matures, virtual manufacturing systems at the component level are already delivering substantial business benefit in specific cases. One example is in the virtual modeling of the casting process (Figure 19). Historically, casting process development has been performed through an expensive 'build and break' approach [12]. Through the use of casting modeling during product and process development, engineers can evaluate new designs for manufacturability and optimize the manufacturing process before real components are cast, during new product introduction or yield improvement activities. By accurately modeling the heat exchange between the casting, mold, and key parts of the furnace, Rolls-Royce has been able to successfully predict porosity, stray grains, and freckle chain defects. Improvements proposed and rapidly tested by iterating the model can then be validated by selected and highly targeted physical trials, significantly reducing the cost and lead time associated with build-and-break-type process development efforts. Investment casting assemblies now typically contain features that would likely not have been developed without the use of modeling. Rolls-Royce has realized significant business benefit from deployment of process modeling systems to casting facilities and suppliers, which facilitate the use of ICME tools by engineers without requiring a high level of ICME knowledge or skill. Such deployed application of modeling has been applied to modify tooling, process, and product designs to deliver significant cost reduction through reduction of non-conformance and improvements in Right First Time production.

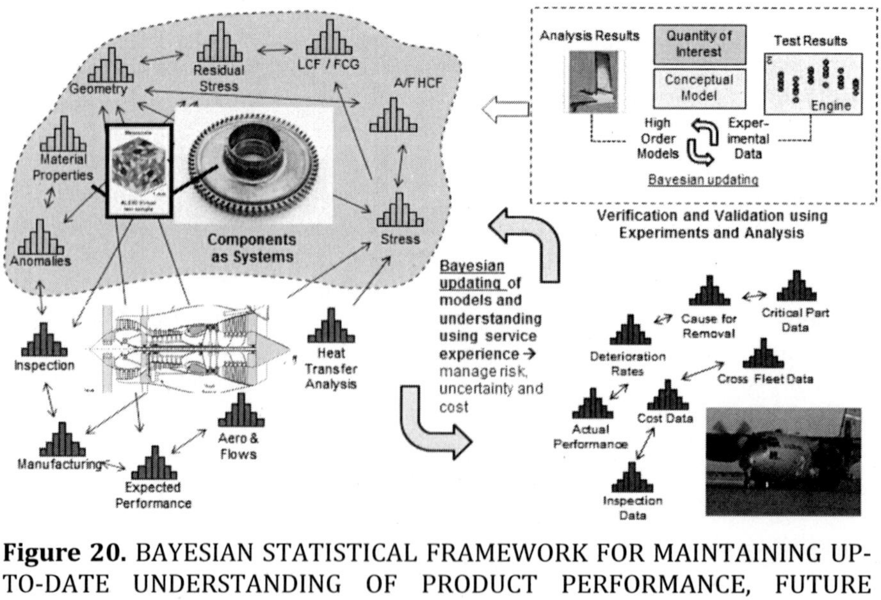

Figure 20. BAYESIAN STATISTICAL FRAMEWORK FOR MAINTAINING UP-TO-DATE UNDERSTANDING OF PRODUCT PERFORMANCE, FUTURE DEVELOPMENT MODELS, AND LIFE CYCLE COSTS.

Life prediction and management for life-limited components

While ICME offers powerful new predictive tools, it also embraces the need to quantify uncertainty in predictive capability as seen in the preceding examples. It follows, from a business perspective, the NRC's vision of ICME as a holistic system that brings together design optimization and manufacturing that encourages further consideration of uncertainties throughout the entire life cycle of a product, not just during development. This is partly due to the nature of optimization and partly to do with the role manufacturing plays in life cycle cost reduction (see Figure 20).

While the objective functions for system design optimization are generally well understood, this is not always true for system boundary conditions. Often, engineers' understanding of boundary conditions improves as more data is acquired from manufacturing and the product's usage in service. For the same reason, it may become apparent as a product matures that it satisfies intended requirements but is no longer optimized for the conditions to which it is actually exposed. Further, as such information becomes apparent, tools like ICME will likely be called upon to help improve performance and cost. There are many possible paths toward product improvement and cost reduction to explore. Early use of engineering expertise in ICME along with later in-service data to update models and prior understanding means that ICME fits well within a broader context of Bayesian frameworks and UQ-based decision analysis [13, 14]. Perhaps one of the best examples in the gas turbine industry is in life prediction and management of high-energy discs.

In application to high-energy discs, computational materials science tools must be carefully linked with many other UQ and decision-making tools and data sources to enable the NRC vision [7]. Design optimization and field management of these components require application of a number of complex, inter-related tools, technologies, physics-based models, cost and reliability models, as well as expert decisions. Moreover, referring back to Figure 9, there are many sources of uncertainty associated with discs. Some of these can be quantified early by ICME's development models such as microstructural characterization, forging, and finished part residual stresses and material properties. Other uncertainties such as operational usage and typical engine performance and deterioration effects must be assumed prior to certification and production. However, in both cases, what are commonly known as Bayesian prior distributions must be assumed. In the case of usage, distributions associated with environment such as ambient conditions or distributions reflecting uncertainty in operator behavior such as pilot decisions to use lower power flex takeoff conditions may be defined. In the case of material uncertainties, the *supply chain* processing parameters and manufacturing target dimensions may be considered Bayesian priors. Furthermore, these two general sources of uncertainty, developmental and in-service, must be updated and managed jointly in order for safety, performance risks, and costs to be effectively managed throughout the life cycle of the product. This can easily be described by way of a simple example if one considers the correlated effects of inelastic material deformation from speed and temperature with location-specific material characteristics such as yield strength. Bayesian posterior distributions, e.g., LCF life, will therefore be functions of *both* material properties and *usage*. As described earlier, ICME then becomes a key part of continual validation and verification. Measureable data such as forging solution temperature, distortion from machining, finished part geometry, and engine air temperatures can be used to update corresponding distributions of quantities that are more difficult to measure in production such as component residual stresses, metal temperatures, and disc stresses. This process of Bayesian updating can be used to generate highly correlated distributions such as disc creep life and damage tolerance capability that naturally follow out of similar prior understandings and data. Furthermore, UQ tools that lend themselves to quantifying system behavior such as *Bayesian networks* can be used to maintain an updated understanding of these distributions [14]. For discs (and other components), these tools are envisioned to work alongside standard statistical and *robust design* methods to link ICME into a holistic production system.

Many UQ tools themselves, like Bayesian networks, will be largely invisible to the supply chain. But the outputs, questions, and decisions associated with these networks will be immediately visible and will be easy to explain in terms both an OEM and supplier will understand. In this way, both parties can supply and discuss implications to both non-proprietary and proprietary models and information from new data as it becomes available. The basic Bayesian reasoning process includes simple cause and effect relationships, prior distributions from expert knowledge and data, likelihood models, and approach to updating understanding. This makes it a natural tool for engineers developing complex systems in the context of ICME. ICME models are envisioned as a powerful new way to link design and the manufacturing supply chain. But as with any model, its ultimate utility is seen in how it is used by those who make business risk-related decisions.

CONCLUSIONS

Significant work has been done across the industry to develop computational models of materials, manufacturing, and design processes; however, there has been less success at integrating these models into product development processes. Application of these models has delivered both development and production benefits, but mostly in isolated one-off situations. The integration of ICME into design and manufacturing systems, though clearly more challenging, promises order of magnitude benefits in cost and lead time reduction. This review has illustrated and emphasized that integration challenges are not just digital but physical, heuristic, and relational as well. To be successful, ICME initiatives must deliver solutions the entire supply chain understands, trusts, and can use. A review of some of the specific challenges faced by ICME users across the supply chain was presented. Ultimately, the transition from technology push to technology pull for use of ICME across the supply chain will reside in the benefit realized through successful use of the technology to deliver product benefit. In response to already realized business benefit on both legacy and new product introduction, Rolls-Royce is integrating ICME into our product development processes as part of our standard methodology going forward.

ACKNOWLEDGEMENTS

Authors would like to express appreciation to Dr. Rollie Dutton and Dr. David Furrer who initially invited a presentation of this work under the same title which was presented during a symposia on 'ICME - from the customer's view point' that Dr. Dutton and Dr. Furrer co-chaired during the 2012 Materials Science & Technology conference in Pittsburgh, PA. Authors would also like to acknowledge contributions from Ben Saunders of Rolls-Royce who supported with images for figure generation and Tony Phipps of Rolls-Royce who supported business benefit articulation. Finally, it is prudent to recognize the purposeful integration efforts being coordinated across several professional societies including TMS, ASM International and AIAA. The lessons learned reported in this paper are collated in part from cross-industry and academia collaboration and feedback regarding the ICME integration efforts of these communities.

AUTHORS' CONTRIBUTIONS

JM conceived, organized, and integrated all individual author contributions to draft the review. AB contributed to materials design content. PB, DH, and KM contributed to virtual manufacturing content. RC contributed to preliminary design content. NC contributed to detailed design content. CD contributed to the detailed design and life prediction and management content. AK contributed to virtual engine content. GM contributed to uncertainty quantification and robust design content. JS contributed to uncertainty quantification, Bayesian methodology and life prediction and management content. All authors read and approved the final manuscript.

REFERENCES

1. Matlik JF, Bolcavage A: Integrated computational materials engineering from a gas turbine engine perspective. In Symposium of integrated computational engineering - the customer's point of view. Pittsburgh, PA: Presentation at the Materials Science & Technology 2012 Conference & Exhibition,; 2012. 8 October 2012 8 October 2012
2. Glavicic MG, Goetz RL, Shen G, Rasche JA, Ress RA III, Stillinger J: Application of ICME to turbine engine component design optimization. In 52nd AIAA/ASME/ASCE/AHS/ASC structures, structural dynamics and materials conference. Denver; 2011. 4–7 April 2011, AIAA-2011–1738 4–7 April 2011, AIAA-2011-1738
3. National Research Council: Accelerating technology transition: bridging the valley of death for materials and processes in defense systems. The National Academies Press, Washington, DC; 2004.
4. Olson GB: Integrated Computational Materials Design: From Genome to Flight. In 54th AIAA/ASME/ASCE/AHS/ASC structures, structural dynamics and materials conference. Boston; 2013. 8–11 April 2013, AIAA-2013–1847 8–11 April 2013, AIAA-2013-1847
5. Martin R, Evans D: Reducing costs in aircraft: the metals affordability initiative consortium. JOM 2000, 52(3):24–28.
6. Cowles B, Backman B, Dutton R: Verification and validation of ICME methods and models for aerospace applications. Integr Mater Manuf Innov 2012, 1: 2.

7. National Research Council: Integrated computational materials engineering: a transformational discipline for improved competitiveness and national security. The National Academies Press, Washington, DC; 2008.

8. Huron ES, Reed RC, Hardy MC, Mills MJ, Montero RE, Portella PD, Jack T: Polycrystalline γ (Ni)/γ′ (Ni_3A1)-δ (Ni_3Nb) eutectic Ni-base superalloys: chemistry, solidification and microstructure. In Superalloys. Edited by: Xie M, Helmink RC, Tin S. Hoboken, NJ: Wiley; 2012:633–642.

9. Villars P, Okamoto H, Cenzual K (Eds): ASM alloy. ASM International, Phase Diagrams Center, Materials Park, OH; 2007.

10. Shen G, Cooper N, Ottow N, Goetz R, Matlik J: Integration and automation of residual stress and service stress modeling for superalloy component design. In Superalloys 2012. Edited by: Huron ES, Reed RC, Hardy MC, Mills MJ, Montero RE, Portella PD, Telesman J. Hoboken: Wiley; 2012:129–134.

11. Keskin A, Saiz A: An integrated design system for optimization of gas turbine components. In 53rd AIAA/ASME/ASCE/AHS/ASC structures, structural dynamics and materials conference. Honolulu; 2012. 23–26 April 2012, AIAA-2013–1408 23–26 April 2012, AIAA-2013-1408

12. EuroPAM: 16th European conference and exhibition on digital simulation for virtual engineering. Toulouse; 2006. 10–12 October 2006 10–12 October 2006

13. Azaidan M, Mills A, Harrison R: Bayesian framework for aerospace gas turbine engine prognostics. In IEEE aerospace conference. Big Sky; 2013. 2–9 March 2013 2–9 March 2013

14. Mahoney S, Stillinger J: Towards a probabilistic framework for integrated computational materials engineering. In 53rd AIAA/ASME/ASCE/AHS/ASC structures, structural dynamics and materials conference. Honolulu; 2012. 23–26 April 2012. AIAA-2013–1529 23–26 April 2012. AIAA-2013-1529

6
CHAPTER

THERMODYNAMIC PERFORMANCE ANALYSIS OF A GAS TURBINE IN AN EQUATORIAL RAIN FOREST ENVIRONMENT

Barinaadaa Thaddeus Lebele-Alawa, Vining Jo-Appah

Department of Mechanical Engineering, Rivers State University of Science and Technology, Port Harcourt, Nigeria

ABSTRACT

This paper discusses the thermodynamic analysis of a gas turbine power plant located in the equatorial rainforest of southern Nigeria. Steady state monitoring and direct collection of data from the Mk IV Speedtronics system and log books in the control room was performed. The variation of operating conditions (ambient temperature, compressor discharge temperature, turbine inlet temperature, exhaust temperature and fuel mass flow rate) on the performance of gas turbine (thermal efficiency, net power output, heat rate, specific fuel consumption and compressor work) were investigated using various thermodynamic relations and equations. The results show that a degree rise in ambient temperature could be responsible for the following: 1.37% reduction in the net power output, 1.48% increase in power drop, 1.49% reduction in thermal efficiency,

2.16% increase in heat rate, 2.17% increase in specific fuel consumption and 0.3% increase in compressor work. Furthermore the thermal efficiency decreases by 0.006% for 1 kcal/kWh increase in heat rate and the heat transfer in the hot gas part was found to increase by 0.16% for a degree rise in ambient temperature. Also the work reveals that the gas turbine had a huge drop in power due to influence of site parameters in contrast to designed data.

Keywords: Gas-Turbine, Thermal Efficiency, Power, Compressor Work, Specific Fuel Consumption, Heat Rate, Combustion

1. INTRODUCTION

Gas turbines are designed to be highly effective in producing aligned high thrusts. In a gas turbine, atmospheric air is drawn in through an intake duct into the compressor and delivered at a higher pressure to the combustor. This is accomplished by the gas turbine compressor consisting of a cascade of several stages of blades located in radial form on a single axle [1] -[3] .

The power produced by an expansion turbine and consumed by a compressor is proportional to the absolute temperature of the gas passing through the device. Consequently, it is advantageous to operate the expansion turbine at the highest practical temperature consistent with economic materials and internal blade cooling tech- nology and to operate the compressor with inlet air flow at a temperature as low as possible.

Overall efficiency of the gas turbine cycle depends primarily upon the pressure ratio of the compressor. It is important to realize that in the gas turbine the processes of compression, combustion and expansion do not occur in a single component as they occurred in a reciprocating engine. It is well known that the performance can be qualified with respect to its efficiency, power output, and specific fuel consumption as well as work ratio. There are several parameters that affect its performance including the compressor compression ratio, combustion inlet temperature and turbine inlet temperature [4] .

Thermodynamic analysis of the gas and steam turbines at Takoradi thermal power station has been studied [5] . The work deals with the validation of thermodynamic models that was used to evaluate the performance of this plant using both light crude oil (LCO) and natural gas (NG). The model inputs and outputs included the ambient temperature and pressures, compressor pressure ratio, component efficiencies, air flow rate in compressor, mass of fuel, fuel calorific values, compressor power input, pressure of water in the heat recovery steam generator (HRSG), temperature of flue gas at the exit of the HRSG, temperature of water at the pump input, mass flows of steam and cooling water and the net power from both the gas and steam turbines. Mass and energy balance equations were used for the model development. A work on the analysis of the combustion process of a gas turbine to control temperature has been done [6] . The combustion analysis was carried out to determine combustion chamber exhaust temperature and turbine entry temperature. In the work, three basic factors: excess air, compressor inlet temperature and combustion chamber wall temperature were considered to determine their influence on the com- bustion chamber exhaust temperature with a view to controlling it (combustion chamber exhaust temperature) to tolerable limits with respect to the plants performance and component failure. In analyzing the effects on the combustion chamber exhaust temperature, the temperature profile on the liner was taken into consideration and an equation was developed to describe the temperature profile of the liner using a one dimensional approach. The result shows that the combustion chamber temperature rise increases as the air/fuel ratio increases at constant air inlet temperature.

Analytical modeling of performance characteristics of axial flow two-stage turbine engine using pressure losses models and comparing with experimental results has been done [7] . In this work a detailed systematic analysis of two-stage, axial flow turbine by using different losses models and a new suggestion algorithm based on one- dimensional simulation. To predict the performance of two-stage axial flow turbine at both the design and off- design conditions, one-dimensional modeling, being an accurate and fast method for obtaining gas turbine per- formance condition was used.

Also, numerical simulation of performance of an axial turbine first stage has been studied [8] . The study deals with steady state 2/3D viscous flow to obtain information about the performance of a first-stage axial turbine by simulating the flow through the blades. Computational fluid dynamics software was used to solve the RANS equations with the spalart-Allmaras turbulence model. Performance map for the whole range of operation of tur- bine was constructed. The CFD results were compared with the mean line loss model ones, which account for the losses by Denton loss model.

The effect of surface roughness on thermodynamic performance parameter of axial flow compressor has also been studied [9] . This study respectively discusses the effect of surface roughness on performance parameter when surface roughness is constant and linearly distributed. The study chooses NASA rotor 37 as study object. Reverse design method was applied to reconstruct the fouled compressor by combining laser triangulation sensor with compressor fouling test rig and then reconstructed solid model is imported into ANSYS CFX to simulate flow field.

Other performance analyses have been carried out such as performance analysis and components irreversibility of a (25 MW) gas turbine power plant modeled with a spray cooler [10] . In this work the potential benefits of improving the performance of the current gas turbine plant into a more advanced cycle with high efficiency and power output through inlet air cooling were analyzed. In the study, performance characteristics were determined for a set of actual operational conditions including ambient temperature, relative humidity, turbine inlet temperature and pressure ratio. Operating data for Gas turbine unit were collected from the daily turbine control log sheet for a period of two years. The average operating variables and the thermodynamic properties were determined using Engineering Equation Software (EES). The analysis of the plant was divided into different control volumes and performance of the plant was estimated using component wise modeling. Mass and energy conservation laws were applied to each component and the performance of the plant was determined for the simple system (without spray cooler) and for the cooled system (with spray cooler). The irreversibility rate and the efficiencies of the turbine components were evaluated employing Kotasexergy models. The results obtained showed that the use of a spray cooler on the existing gas turbine cycle gives a better thermal efficiency and less irreversibility rate in the component system and the entire plant.

Also thermodynamic performance analysis of gas turbine power plant has been studied [4] . In this work, the presentation of the parametric study of thermodynamic performance on gas turbine power plant was done. The variation of operating conditions (compression ratio, turbine inlet and exhaust temperature, air to fuel ratio, isentropic compression and turbine efficiency, and ambient temperature) on the performance of gas turbine (thermal efficiency, compressor work, power, specific fuel consumption, heat rate) were investigated. The analytical formula for the specific work and the efficiency were derived and analyzed. The results show that the compression ratio, ambient temperature, air to fuel ratio as well as isentropic efficiencies had strong influence on thermal efficiency.

Thermodynamic appraisal of gas turbine performance in the Niger Delta has also been studied [11] . In this work, the procedure for the study included collation of data from service records, field measurements, simulation of systems and conditions and thermodynamic analysis of results. The results show that a degree centigrade rise in ambient temperature could be responsible for the following: 0.83% reduction in power output, 0.17% increase in heat rate and 0.40% decrease in required air flow rate. Fouling which results in a unit reduction in compressor air flow capacity could yield between half and a unit reduction in compressor isentropic efficiency and the trends tend to follow a linear relationship.

2. MATERIALS AND METHODS

The research methodology involved collection of data from February 2009 to April 2013 as follows:

1. Monitoring and collection of data from control room Mk IV Speedtronics system;
2. Direct reading of design values from installation document;
3. Field investigation of gas turbine generator during turn around maintenance (TAM);
4. The use of thermodynamic relations and equations for such phenomena that could not be directly measured;
5. Analysis of results obtained;
6. Discussion of results obtained, conclusions and making recommendations.

Each method of data collection was designed to produce facts about some aspects of the work. The data obtained from monitoring of the control room Mk IV Speedtronics system and the daily log sheets were intended to determine those factors that influence the gas turbine thermodynamic process. Two thermodynamic properties— pressure and temperature—stood out as important, as they were measured at the entries and exits of the major components of the gas turbine; namely the compressor, combustion chamber and turbine.

In the treatment and collection of data, mean values of daily parameters were computed by the use of statistical method; followed by monthly average and the overall average for the research period. Some of the phenomena of the operation of the set could not be investigated directly by field measurements because the points were the measurements would have been made where inaccessible. One of such phenomena was the combustion tempera- ture (T_3) of the hot gas as it passes through the combustion chamber. This could not be measured directly and as such the parameter was obtained by the use of thermodynamic relations and equations.

The Net Power Output (W_{net}) is the power generated by the generator and is given as:

$$W_{\text{net}} = W_t - W_c \tag{1}$$

where W_t is the shaft work of the turbine and is given as:

$$W_t = m_p C_p \left(T_4 - T_3\right) \tag{2}$$

where m_p is mass of product (kg/s), C_p is specific heat capacity of product, $T_3 = TIT =$ turbine inlet temperature. Total heat supplied (Q_{add}), is calculated from the equation:

$$Q_{\text{add}} = \left(m_a + m_f\right) \times Cp_a \left(T_3 - T_2\right) = m_f \times CV \tag{3}$$

$$T_3 = \frac{m_f \times CV}{\left(m_a + m_f\right) \times Cp_a} + T_2 \tag{4}$$

where m_a is mass flow of air, m_f is mass flow of fuel, Cp_a is specific heat capacity of air and CV is the calorific value of the fuel. To get T_3 for a different fixed amount of fuel supply at each T_1, it was therefore necessary to extrapolate.

For steady flow steady state condition, the extrapolation function is given as:

$$ET = CET + \left(ATTI - CET\right)\left(FFS/AFS\right) \tag{5}$$

where

$ET =$ Extrapolated Temperature, °K;

$CET =$ Compressor Exit Temperature, °K;

$ATTI =$ Actual Temperature at Turbine Inlet for Actual Fuel Supply, °K;

$FFS =$ Fixed Fuel Supply, m^3/s;

$AFS =$ Actual Fuel Supply, m³/s.

Exhaust Temperature T_4 can be determined as:

$$T_4 = \frac{T_3}{(P_2/P_1)^{\frac{(\gamma-1)}{\gamma}}} \tag{6}$$

where γ is isentropic index of compression of air. For the two isobaric processes, $P_2 = P_3$ and $P_4 = P_1$. Thus the turbine pressure ratio P_3/P_4 is equal to the compressor pressure ratio, $r = p_2/P_1$. With the turbine inlet temperature T_3 known, exhaust temperature T_4 can be determined from Equation (6) above.

Compressor Work (W_c) of the compressor is given as:

$$W_c = m_a Cp_a (T_2 - T_1) = \rho_a V_a Cp_a (T_2 - T_1) \tag{7}$$

where ρ_a is density of air, V_a is volume of air aspirated by the compressor, T_1 is ambient temperature and T_2 is compressor discharge temperature.

Thermal Efficiency: The gas turbine thermal efficiency (η_{th}) is the percentage of the total fuel energy input that appears as the net work output of the cycle.

$$\eta_{th} = \frac{\text{Net Work}}{\text{Heat Supplied}} \tag{8}$$

where net work is power output and is given as Equation (8) above.

Specific Fuel Consumption (SFC): The ratio of fuel used by a machine to a certain force such as the amount of power the machine produces. And it can be determined by the equation:

$$SFC = \frac{3600 \times mf}{W_{\text{net}}} \tag{9}$$

where $m_f =$ is fuel mass flow rate (kg/s).

The Heat Rate (HR) is a measure used to determine how efficiently a generator uses heat energy. It can be expressed as:

$$HR = \frac{\text{Heat Supplies}}{\text{Power Generated}} \tag{10}$$

Stoichiometric Equation: This is the ideal combustion process in which minimum amount of air (stoichiometric or theoretical air) is needed to completely burn a fuel.

$$C_mH_n + \left(m + \frac{n}{4}\right)O_2 + 3.76\left(m + \frac{n}{4}\right)N_2 \rightarrow mCO_2 + \frac{n}{2}H_2O + 3.76\left(m + \frac{n}{4}\right)N_2 \tag{11}$$

Assuming steady-state, steady-flow process involving ideal gasses, we can write from the first law of ther- modynamics applied to combustion as;

$$Q + \sum_R nRi\left[h_f^o + \Delta h\right] = W + \sum_R nPe\left[h_f^o + \Delta h\right] \tag{12}$$

where

Q = Heat transfer rate in ω;

nRi = Number of moles of each gas constituents of the reactants;

nPe =Number of moles of each gas constituents of the product;

h_f^0 = Heat of formation, in KJ/Kmol;

Δh = Difference in enthalpy between any given state and the enthalpy at reference temperature at which the heat of formation is taken in KJ/Kmol. This is generally taken at 25°C (298 K);

W = Work transfer in ω;

$\sum$ = Summation.

Subscript "P" and "R" indicate "products" and "reactants" respectively.

3. RESULTS AND DISCUSSION

The parameters in Table 1 were obtained from the installation document of MS6001B gas turbine.

The values of the turbine inlet temperature in Table 2 were calculated using Equation (4). The parameter in- fluence in terms of ambient temperature on the performance of gas turbine cycle power plant is presented in Tables 1-9. The effects on the operating conditions on the power output, thermal efficiency, heat rate, specific fuel consumption and compressor work were determined and plots are shown in Figures 1-8. In Table 3 the percentage of design power output is gotten by comparing actual power output with the value of the design power output.

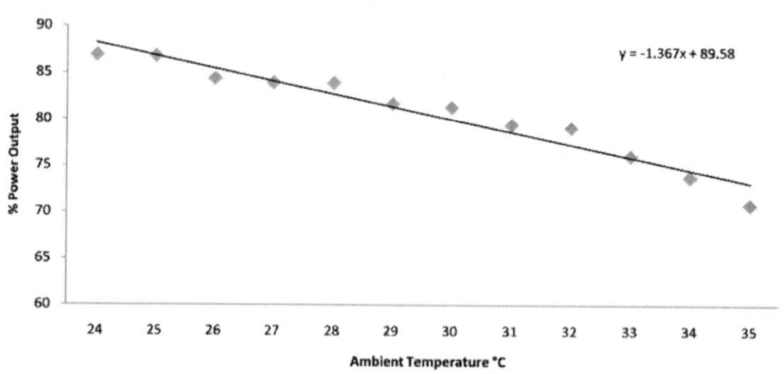

Figure 1. Effect of ambient temperature on power output.

Table 1. Design data.

S/NO	Parameter	Units	Design Data
1	Power Output	MW	38.12
2	Thermal Efficiency (η_{th})	%	28.3
3	Heat Rate (HR)	kcal/kW·h	3047
4	Specific Fuel Consumption (SFC)	kg/kW·h	0.283
5	Compressor Work	MW	45.28

Table 2. Averages of temperature, flow rate and power output (measured).

Ambient Temperature	Compressor Exit Temperature	Fuel Gas Supply	Turbine Inlet Temperature	Exhaust Temperature	Power Output
°C (T_1)	°C (T_2)	kg/s	°C (T_3)	°C (T_4)	MW
24	340	3	1217	532	33
25	341	3	1218	533	33
26	343	3	1220	534	32
27	345	3	1222	535	32
28	348	3	1225	536	32
29	349	3	1226	537	32
30	350	2.9	1199	522	30
31	353	2.8	1173	519	28
32	355	2.5	1088	462	26
33	357	2.4	1062	459	26
34	359	2.3	1036	434	24
35	360	2.3	1036	434	24

Table 3. Ambient temperature and turbine inlet temperature and power output.

Ambient Temperature	Compressor Exit Temperature	Extrapolated Turbine Inlet Temperature	Exhaust Temperature	Actual Power Output	Actual Power Output ×100 / Design Power Output	Power Drop
°C (T_1)	°C (T_2)	°C (T_3)	°C (T_4)	MW	%	%
24	340	1217	532	33.14	89.94	10.06
25	341	1218	533	33.10	86.83	13.17
26	343	1220	534	32.18	84.42	15.58
27	345	1222	535	32.00	83.95	16.05
28	348	1225	536	32.00	83.95	16.05
29	349	1226	537	31.15	81.72	18.28
30	350	1228	538	31.00	81.32	18.68
31	353	1232	540	30.27	79.41	20.59
32	355	1235	542	30.16	79.12	20.88
33	357	1238	544	29.00	76.08	23.92
34	359	1241	545	28.13	73.79	26.21
35	360	1242	546	27.00	70.83	29.17

Table 4. Ambient temperature on thermal efficiency.

Ambient Temperature	Actual η_{th}	Actual × 100 Design Data
°C (T_1)	%	%
24	24.60	86.93
25	24.50	86.57
26	24.00	84.81
27	23.90	84.45
28	23.70	83.75
29	23.00	81.27
30	22.30	78.80
31	22.00	77.74
32	22.00	77.74
33	21.00	77.20
34	20.80	73.50
35	20.00	70.67

Table 5. Ambient temperature on heat rate.

Ambient Temperature °C (T_1)	Actual HR kcal/kW·h	Actual HR × 100 Design Data %
24	3508	115.13
25	3508	115.13
26	3626	119.00
27	3630	119.13
28	3630	119.13
29	3688	121.04
30	3748	123.01
31	3837	125.93
32	3851	126.39
33	4009	131.57
34	4132	135.61
35	4303	141.22

Table 6. Ambient temperature on specific fuel consumption.

Ambient Temperature °C (T_1)	Actual SFC kg/kW·h	Actual SFC × 100 Design Data %
24	0.326	115.19
25	0.326	115.19
26	0.336	118.73
27	0.337	119.08
28	0.337	119.08
29	0.343	121.20
30	0.348	122.97
31	0.357	126.15
32	0.358	126.50
33	0.372	131.45
34	0.384	135.69
35	0.400	141.34

Table 7. Ambient temperature on compressor work.

Ambient Temperature °C (T_1)	Actual W_c MW	Actual W_c × 100 Design Data %
24	47.54	104.99
25	47.54	104.99
26	47.69	105.32
27	47.84	105.65
28	48.14	106.31
29	48.14	106.31
30	49.17	108.59
31	49.47	109.25
32	49.50	109.32
33	49.62	109.58
34	49.84	110.07
35	49.90	110.20

Table 8. Heat rate on thermal efficiency.

Ambient Temperature °C (T_1)	Heat Rate kcal/kW·h	Thermal Efficiency %
24	3508	24.60
25	3508	24.50
26	3626	24.00
27	3630	23.90
28	3630	23.70
29	3688	23.00
30	3748	22.30
31	3837	22.00
32	3851	22.00
33	4009	21.00
34	4132	20.80
35	4303	20.00

Table 9. Ambient temperature and turbine inlet temperature and heat transfer.

Ambient Temperature	Compressor Exit Temperature	Extrapolated Turbine Inlet Temperature	Actual Heat Transfer	Percentage Heat Transfer
°C (T_1)	°C (T_2)	°C (T_3)	kJ/kmol	%
24	340	1217	637,824	120.87
25	341	1218	640,462	121.37
26	343	1220	640,989	121.47
27	345	1222	641,517	121.57
28	348	1225	642,045	121.67
29	349	1226	642,572	121.77
30	350	1228	643,100	121.87
31	353	1232	644,156	122.07
32	355	1235	644,683	122.17
33	357	1238	645,739	122.37
34	359	1241	647,322	122.67
35	360	1242	649,433	123.07

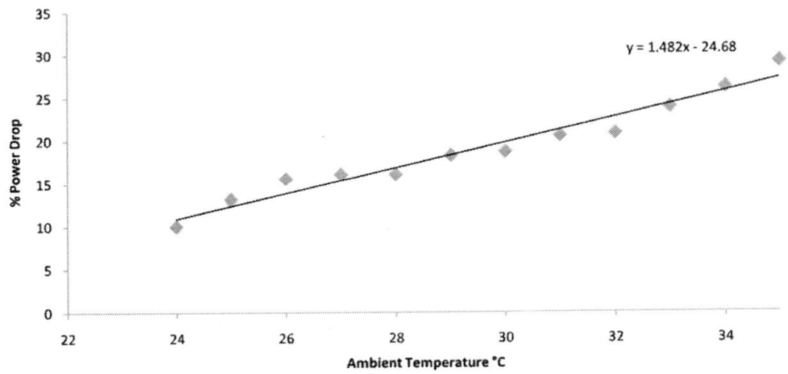

Figure 2. Effect of ambient temperature on power drop.

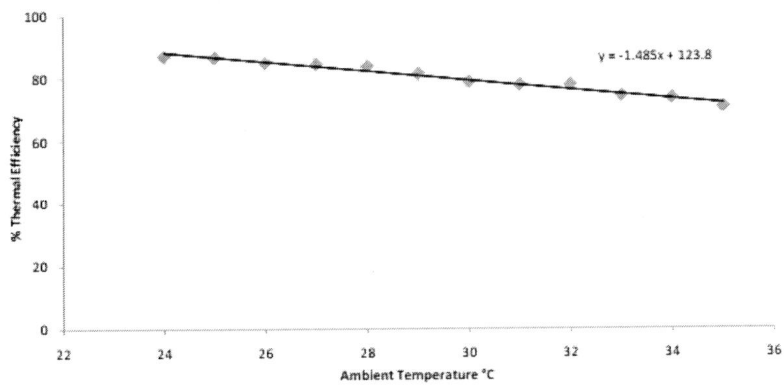

Figure 3. Effect of ambient temperature on thermal efficiency.

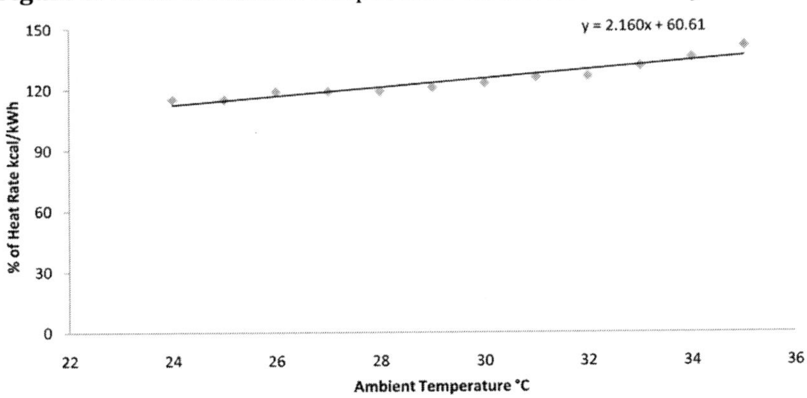

Figure 4. Effect of ambient temperature on heat rate.

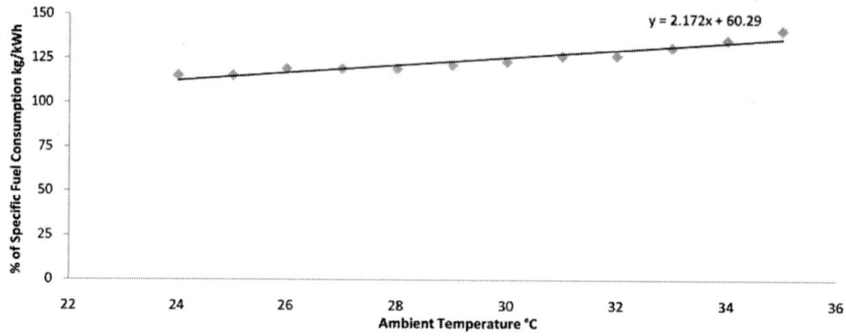

Figure 5. Effect of ambient temperature on specific fuel consumption.

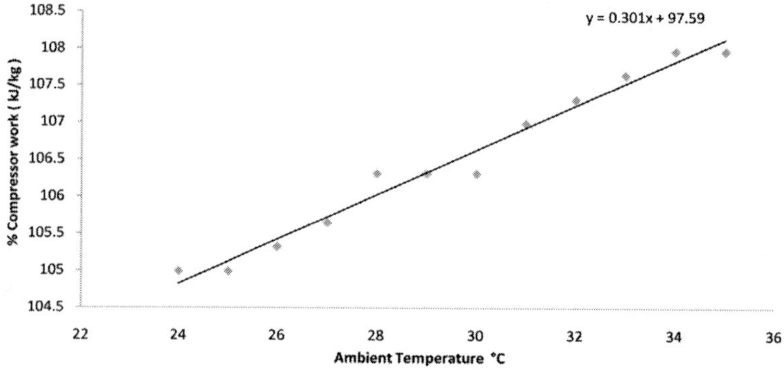

Figure 6. Effect of ambient temperature on compressor work.

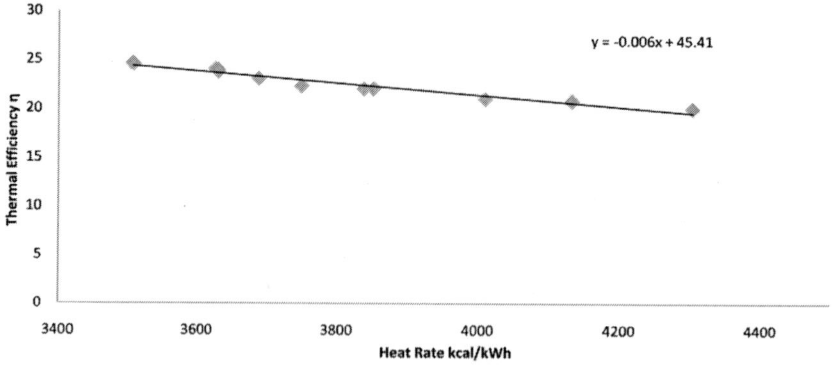

Figure 7. Effect of heat rate on thermal efficiency.

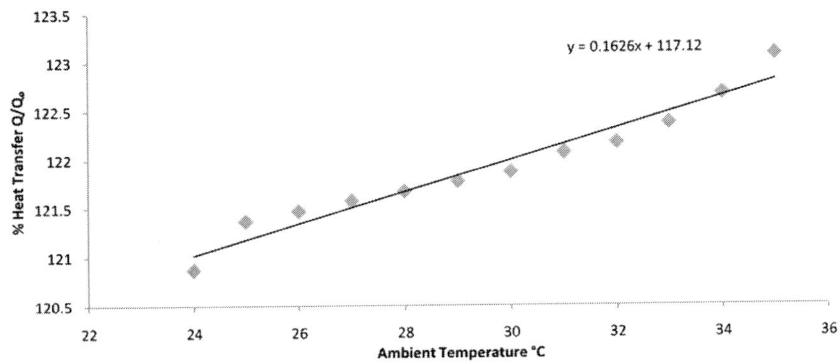

Figure 8. Effect of ambient temperature on heat transfer.

The governing equation is

$$y = -1.3676x + 89.586 \Rightarrow \frac{\delta y}{\delta x} = -1.3676$$

Based on Figure 1 and the resultant straight line equation above, it shows that there is a fall in the power output of about 1.37% for every 1°C rise in the ambient temperature. Translated in real terms it means if a gas turbine operated at an average ambient temperature of 30°C instead of 15°C used for the design, there will be a fall in power output of 20.55% or 7.83 Mw for a gas turbine design to generate 38.12 MW at 15°C.

Based on Figure 2 and the resultant straight line equation above, it shows that there is a rise in the drop of the power output of about 1.48% for every 1°C rise in the ambient temperature. Translated in real terms it means if a gas turbine operated at an average ambient temperature of 30°C instead of 15°C used for the design, there will be a rise in the power drop of 22.2% or 8.46 MW for a gas turbine design to generate 38.12 MW at 15°C.

For this reason colossal losses are incurred in power generating industries in the tropical regions. To have an idea, let us consider some of the gas turbine power generating sets in the Niger Delta.

Figure 3 above shows that there is a fall in thermal efficiency of about 1.49% for every 1°C rise in the ambient temperature. Translated in real terms it means that if a turbine set operated at an average ambient temperature of 30°C instead of 15°C used for the design, there will be a fall in thermal efficiency of 22.35% or thermal efficiency of 6.33% for a gas turbine designed to produce 28.3% η_{th} at 15°C.

Figure 4 shows that there is a rise in heat rate of about 2.16% for every 1°C rise in the ambient temperature. Translated in real terms it means that if a turbine set operated at an average ambient temperature of 30°C instead of 15°C used for the design, there will be a rise in heat rate of 32.4% or 987.23 kcal/kW×h for a gas turbine designed to generate 3047 kcal/kW×h at 15°C.

Figure 5 shows that there is a rise in specific fuel consumption of about 2.17% for every 1°C rise in the ambient temperature. Translated in real terms it means that if a turbine set operated at an average ambient temperature of 30°C instead of 15°C used for the design, there will be a rise in specific fuel consumption of 32.55% or 0.092 kg/kWh of fuel for a gas turbine designed to have a specific fuel consumption of 0.283 kg/kWh at 15°C.

Figure 6 shows that there is an increase in compressor work of about 0.3% for every 1°C rise in the ambient temperature. Translated in real terms it means that if a turbine set operated at an average ambient temperature of 30°C instead of 15°C used for the design, there will be an increase in compressor work of 4.5% or 2 MW for a gas turbine designed to have a compressor work of 45.28 MW at 15°C.

Figure 7 shows that there is a decrease in thermal efficiency of about 0.006% for every rise in heat rate of about 1 kcal/kW×h. Translated in real terms it means that if a turbine set operated at an average ambient temperature of 30°C having heat rate of about 3748 kca/kW×h instead of 15°C as 3047 kcal/kW×h as design, there will be a decrease in thermal efficiency of 22.5%.

Figure 8 shows that there is an increase in the heat transfer of about 0.16% for every rise in ambient temperature. Translated in real terms it means that if a turbine set operated at an average ambient temperature of 30°C instead of 15°C used for the design, there will be a rise in the heat transfer of about 2.4%. This in effect means higher temperature in the hot gas part. And where there is insufficient air mass flow in the inlet due to choking of inlet air filters as a result of high dust content in the Niger Delta area, the consequence will be undercooling in the hot gas part.

4. CONCLUSIONS

It may be pertinent to state here that though much of the data collected were obtained from the performance records of the MS6001B type gas turbine, the findings can be extended to other types of gas turbines perharps with some little modifications. This is because the processes involved in any open gas turbine are the same—a compression process an approximately constant pressure addition of heat and an expansion process. These pro- cesses also take place in standard equipment namely, compressor, combustion chamber and turbine.

From the data analyses and results obtained, it can be generally said that the climatic condition that is peculiar in the site that was not fully addressed at the time of installation of the gas turbine affected the operations and performance of the set. This particular condition viewed by the author is the high ambient temperature of the area which has a mean daily value of about 30°C and which varies only slightly on both sides of this value as against the general design ambient temperature of 15°C. This high ambient temperature is a negative factor and it affects the thermodynamic process of compression, addition of heat and expansion. Apart from affecting the processes, the components in which these processes do occur namely the compressor, the combustion chamber and the turbine can also be physically affected. Petrochemical gas turbine was chosen as a case study for the analysis because it is in the Niger Delta area where all the known larger power generating gas turbines in the country are installed and it is expected that this will not change for some time to come because the fuel for the combustion is abundant in the area. With the foregoing, the gas turbine seems to be the best option for large scale generation of power in Nigeria. Nevertheless, huge amount of loss in power is experienced due to high ambient temperature. Gas turbines are the power of industrial plants in which they are installed in terms of electricity generation within the plant. And as such in order to maintain them at high efficient level. A periodic performance evaluation of the equipment is required.

This work reveals the various parameter and equations used in assessing the performance of the machine and also to check the deterioration of the gas turbine design operation with time by steady monitoring, direct collection of data and using the various thermodynamic relations and equations. The use of micro soft excel is used to obtain the summary of the overall results. The results were compared with the equipment design data. It provides a good method of obtaining the performance of gas turbine.

REFERENCES

1. Lebele-Alawa, B.T., Hart, H.I., Ogagi, S.O.T. and Probert, S.D. (2008) Rotor-Blades' Profile Influence on a Gas Turbine's Compressor Effectiveness. Applied Energy, 85, 494-505.
2. Lebele-Alawa, B.T. (2010) Axial Thrust Responses to Gas Turbine's Rotor-Blade Distortions. Journal of Engineering Physics and Thermophysics, 83, 991-994.
3. Cohen, H., Rogers, G.F.C. and Saravanamuttoo, H.I.H. (1998) Gas Turbine Theory. 4th Edition, Addison Wesley Long- man Ltd., London.
4. Rahman, M.M., Thamir, K.I. and Ahmed, N.A. (2012) Thermodynamic Performance Analysis of Gas Turbines Power Plant. Journal of the Physical Sciences, 6, 3539-3550.
5. Charles, M., Abeeku, B.H. and Seth, P.A. (2010) Thermodynamic Analysis of the Gas and Steam Turbine at Takoradi Thermal Power Station. Journal of Technology and Advanced Engineering Research, 1, 62-72.
6. Odubo, E.O. (1997) Air Flow Regulator for Gas Turbine Combustion Chamber Temperature Control. Master of Technology Dissertation, Rivers State University of Science and Technology, Nigeria.
7. Javaniyan, J., Eftari, M., Kaliji, H.D., Ghadak, F. and Rad, M. (2013) Analytical Modeling of Performance Characteristics of Axial Flow Two-Stage Turbine Engine Using Pressure Losses and Comparing with Experimental Results. Journal of Applied Science, 21, 1250-1259.
8. Vinicius, G.M., Edson, L.Z., Claudia, R.A. and Lima, R.C. (2012) Numerical Simulation of Performance of an Axial Turbine First Stage. Journal of Aerospace Technology, 4, 175-184.
9. Huadong, Y. and Hong, X. (2013) The Effect of Surface Roughness on Thermodynamic Performance Parameters of Axial Flow Compressor. Journal of Applied Science, Engineering and Technology, 5, 4458-4463.
10. Fidelis, I.A., Ikpi, U.U. and Dodoye, I.I. (2012) Performance Analysis and Components Irreversibities of a (25 MW) Gasturbine Power Plant Moduled with a Spray Cooler. Journal of Engineering and Applied Services, 5, 35-41.

11. Hart, H.I. (1998) Thermodynamic "Appraisal of Niger Delta Gas Turbine Performance". Ph.D. Dissertation, Mechanical Engineering, University of Nigeria, Nsuka.

7

CHAPTER

THE INFLUENCE OF INLET AIR COOLING AND AFTERBURNING ON GAS TURBINE COGENERATION GROUPS PERFORMANCE

Ene Barbu[1], Valeriu Vilag[1], Jeni Popescu[1], Bogdan Gherman[1], Andreea Petcu[1], Romulus Petcu[1], Valentin Silivestru[2], Tudor Prisecaru[1], Mihaiella Cretu[1] and Daniel Olaru[1]

[1] *INCDT COMOTI, Bucharest, Romania*

[2] *Politehnica University, Bucharest, Romania*

1. INTRODUCTION

Usually, cogeneration is defined as combined production of power and thermal energy from the same fuel source, represented by natural gas, liquid fuel, refinery gas, etc. In conventional energy production the efficiency is approximately 40 %, but through cogeneration it can reach even 90 %. Fuel supply and increased performance requirements, environment concerns, continuously variable market conditions have contributed to the development of the gas turbines. The performances, exploitation costs, safety in operating conditions have made these installations to be selected for cogeneration processes.

2. STATE OF ART

Gas turbine systems operate on the ideal thermodynamic cycle (consisting in two isentropic and two isobars) represented by Brayton cycle. The real Brayton cycle consists in quasiadiabatic expansion and compression processes, but unisentropic, and the heat transfer processes are not isobar processes, due to flow pressure losses. In addition, the air and hot gases are not perfect gases and not have the same flow rates. Brayton cycle thermal efficiency depends on: compression ratio; ambient temperature; air temperature at turbine inlet; compressor efficiency and turbine components efficiency; blade cooling requirements; increased performance systems (exhaust gases heat recovery, intercooling, intake air cooling, afterburning implementation, fluids injection – water/steam, etc.). The main parameters that define the operating thermodynamic cycle of gas turbine installations (usually disclosed by the suppliers in catalogues) are the temperature at the gas turbine inlet (T_3) and the compression ratio. Generally, gas turbine manufacturers declare performances without taking into consideration the inlet and outlet pressure losses. Gas turbine installations performances are affected by the variation of these parameters as follows [2]: temperature increase at the gas turbine inlet leads to an increase in power and efficiency; the efficiency becomes maximum at a given value of the compression ratio (in T_3=const. hypothesis); there is a value of the compression ratio for which the power is maximum (T_3 and compressor intake air flow rate remains constant). The productivity of a gas turbine cogeneration group depends on the quantity of heat recovered from the turbine exhaust gases (approximately 60-70 % from the fuel energy). This is achieved by adding a heat recovery steam generator in order to supply hot water or steam. Determining factors in total efficiency of the cogeneration group are the gas turbine outlet temperature and the temperature at the stack of the heat recovery steam generator. The combination temperature at the gas turbine inlet – compression ratio determins the outlet temperature. The gas turbine, being located at the upstream of the heat recovery steam generator, significantly influences the cogeneration group performances. The air is induced by the gas turbine compressor in ambient conditions imposed by the location of the cogeneration plant. Compressor inlet temperature and intake air density dictates mechanical work required by the compression process, the fuel and quantity of fuel to be used in order to obtain the necessary temperature at the gas turbine inlet (T_3). Consequently, output power, efficiency, exhaust gases mass flow and outlet temperature (respectively the quantity of heat recovered) are influenced by ambient conditions [3]. The location of the gas turbine cogeneration plant imposes climatic conditions and requires adequate technical solutions in order to ensure performances. Generally, for cogenerative applications, the gas turbine is designed to operate in standard conditions, established by the International Standards Organization and defined as ISO conditions: 15 ^{0}C, 1.013 bar and 60 % humidity. During summer season air temperature rises and its density decreases, leading to a decrease in the intake air mass flow; consequently decreases and power output because it is proportional to the intake air mass flow rate. Without taking supplementary measures, both gas turbine output power and efficiency drop. In the scientific literature there are various papers that deal with the gas turbine's performance dependence of the intake air temperature variation [3-10]. In [4] it is shown that: an increase of 10 ^{0}C at the compressor inlet reduces the gas turbine outlet power with 18%; in comparison with the operation during winter season, the increase of ambient temperature leads to a decrease in gas turbine plants power output with 25-35%, also leading to an average increase of the consumption of 6%. The effect of intake air temperature over the performances differs from one gas turbine to another, but, generally, aeroderivative gas turbines are more sensitive to this phenomenon than the industrial gas turbines [5]. During summer season, when the days are long and hot, the power requirements increase for the residential spaces ventilation, offices, store rooms, etc. Additional energy consumption can be ensured by starting other backup groups, or compensating the loss of power through various other methods. The usual compensation methods of power loss are [6, 7]: compressor inlet air cooling (pre-cooling), intermediate cooling (intercooling), using recovery cycle. Mainly there are two basic compressor inlet air cooling methods: evaporative cooling (with evaporative media cooling or water injection in the inlet air-fogging); refrigeration system cooling [8]. For a 79 MW gas turbine, equipped with a fogging cooling system, the researches conducted at Mashhad (in Iran) showed that during a day, the maximum increase in power is achieved in the afternoon, when the temperature is higher and relative humidity is lower [9]. Inlet air cooling systems analysis in order to be applied to a gas turbine V94.2, in terms of efficiency increase, led to the conclusion that the fogging cooling system meets the design requirements and leads to an increase in power output of approximately 6 MW [10]. With the help of GT PRO software the performances of a 100 MW gas turbine model were analyzed, for a various types of inlet air cooling systems, and it had been reached that a decrease of air temperature of 1 ^{0}C (in the 25-35 ^{0}C interval) leads to an increase in power output of approximately 0.7 MW [11]. For a gas turbine cycle, with intermediate cooling (intercooler), the decrease of inlet air temperature causes the output power to rise and the intercooling leads to a 5-9% gain of power and a 8% reduction in fuel consumption [12]. Reduction of fuel consumption represents a priority both for industrial gas turbine manufacturers and also for the civil aviation. The search is on for new materials that meet the requirements imposed by the higher strains of the gas turbines [13] and also the development of new technologies, including technological transfer from aviation domain to power generation domain. Thus, in aviation, afterburning is used in order to increase traction of supersonic engines. The introduction of afterburning into cogenerative applications leads to an increase in flexibility and global efficiency of the cogeneration group. Afterburning application is possible due to the fact that exhaust gases at turbine outlet have a 11-16% (volumes) content of oxygen [14]. The afterburning installation, located between the gas turbine and the heat recovery steam generator, interacts with the gas turbine but influences the heat recovery steam generator operation especially [14, 15]. To increase the performance of gas turbine cogeneration groups research focused specifically on [16]: increase in burning temperature; increase in compression ratio; improving the methods of design, cooling and burning technologies, and also advanced materials; technological transfer from the aviation domain in the industrial gas turbine domain and conversion of aviation gas turbine (with outdated lifetime) to energy conversion; integrated systems (combined cycles, compressor inlet air cooling, intercooling, turbine exhaust gases heat recovery, afterburning implementation, chemical recovery, etc.). Following the direction displayed in the field, the chapter integrates data from scientific literature with research developed at INCDT COMOTI Bucharest, regarding gas turbine inlet air cooling and afterburning application, as base methods for increasing performances and flexibility of cogenerative group.

3. INFLUENCE FACTORS AND METHODS OF INCREASING PERFORMANCES IN GAS TURBINE COGENERATIVE GROUPS

For a combined cycle (considering as variables ambient temperature, gas turbine outlet temperature and stack temperature) it is shown that the dominant factor in global efficiency rise is stack temperature [17]. Obtaining a high efficiency involves the optimization of the entire cogenerative plant (gas turbine, afterburning installation, heat recovery steam generator, etc.). The efficiency must be maintained even at partial loads (even under 50%) in variable conditions modification. In general, although the target is obtaining a maximum efficiency, nevertheless an adequate flexibility to process requirements is desired, the afterburning installation contributing to this.

3.1. Influence factors

Ambient parameters (humidity, pressure, temperature) can vary significantly depending on geographic location and season, affecting air density and implicitly the gas turbine cogenerative group performances. In the past, the effect of air humidity was neglected but the increase in gas turbine cogenerative groups power and the introduction of water/steam in the combustion chamber made this effect to be reconsidered. Thus, some authors [18] consider that air relative humidity (even at temperatures higher than 10 ^{0}C) has a neglectable influence over the gas turbine output power (as the other performance parameters). This leads to the fact that in some calculus (especially when the results are presented in correlation to ISO conditions) the variations in atmospheric humidity and pressure to be neglected. Others consider that due to the fact that water content modifies thermodynamic properties of inlet air (density, specific heat), at certain gas turbines (depending on specific processes) the performances may increase when humidity rises and in the case of some gas turbines the performances may decrease in the same conditions [19]. However, the increase in relative humidity leads to a significant reduction of NO_x emissions [20].

Ambient pressure is defined by the conditions from plant location, altitude modification leading to air density modification and implicitly to power output variation. Thus, 3-4% losses occur for each 304.8 m (1000 ft) rise in altitude [21].

Power and efficiency of the gas turbine group decrease along with ambient temperature, in figure 1 linear approximate variations being presented. Specific fuel consumption increases with the ambient temperature rise [22].

Gas turbines operate on a wide variety of gaseous fuels (natural gas, liquefied natural gas-LNG, liquid petroleum gas-LPG, refinery gas, etc.) and liquid fuels (kerosene. no. 2 diesel, jet A, etc.). Using a certain type of fuel for the gas turbine has a profound impact both on the design and also on material selection. Usage of liquid fuels imposes: ensuring burning without incandescent particles and residues on the combustor and turbine; reducing hot gas corrosive effect due to aggressive compounds (sulphur, led, sodium, vanadium, etc.); resolving pumping and pulverization (filtering, heating, etc.) issues. In case of using gaseous fuels, a simpler solution is presented due to their higher thermal stability, higher heating power, lack of ash and smut. However, in order to ensure pressure level (required by the gas turbine, afterburning installation, etc.), water and various impurities elimination implies a control-measuring station for the gaseous fuels used (natural gas in the case of cogenerative plant 2xST18 – Figure 2). Although the main fuel for the operation of gas turbine cogenerative groups is natural gas, the economic rise and environmental requirements issued an alternative. The gas turbine can be designed to operate on a variety of fuels, but the rapid transition to other fuel operation, without machine damage or exceeding the level of emissions, still remains an issue subjected to study.

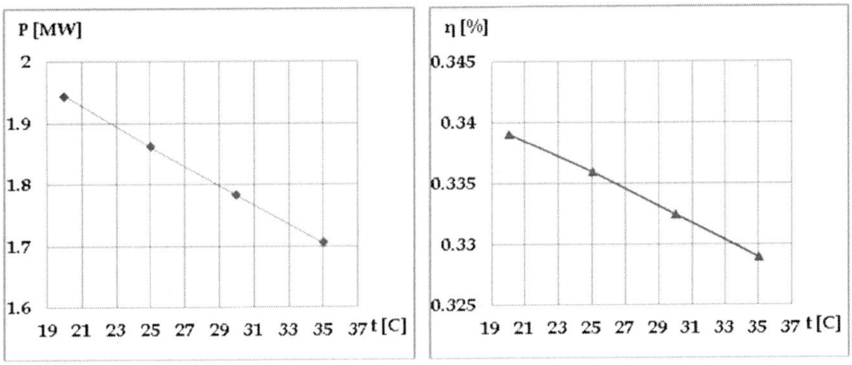

Figure 1. Power output (left) and efficiency (right) variations versus ambient temperature, at a simple gas turbine cycle

Interchangeability at gaseous fuel gas turbine cogenerative groups, represents the ability to change a fuel with another one, without affecting the application or the equipment in which the gaseous fuel is burned [15]. At a constant fuel composition (in the case that the holes through which the fuel passes to the burner have fix dimensions), the quantity of heat delivered by the burner is proportional with the mass flow and heating power. When the composition varies and it is the problem of replacing a fuel with another equivalent, Wobbe index (after John Wobbe-engineer and mathematician) is used as comparison criteria. It is defined as ratio of the lower heating value (LHV) and the square root of relative fuel density respectively, in relation to air (d_{rel}):

$$Wo = LHV / (d_{rel})^{0.5} \tag{1}$$

$$d_{rel} = \rho_{comb} / \rho_{air} \tag{2}$$

Thus, two gaseous fuels (with different chemical compositions), with the same Wobbe index, are interchangeable and the quantity of heat delivered to the equipment is equivalent at the same fuel supply pressure. In order to take into account the fuel temperature, a Wobbe temperature corrected index can be used. According to [23], two gaseous fuels are interchageable if the following relation is satisfied:

$$\frac{\Delta p_2}{\Delta p_1} = \left(\frac{Wo_2}{Wo_1}\right)^2 \left(\frac{A_1}{A_2}\right)^2 \tag{3}$$

where Δp_1 and Δp_2 represent the overpressures of gas 1 and gas 2 respectively, Wo_1 and Wo_2 – Wobbe indices of fuels 1 and 2, A_1 and A_2 – gaseous fuel injection nozzle surface areas.

Figure 2. Cogenerative plant 2xST 18 – Suplacu de Barcau (left) and afterburning installation (right) [14, 15]

Thus, the validation criteria of adequate replacement of one fuel with another equivalent fuel are given by: self-ignition, flame temperature (with a high influence on NO_x emissions forming), flame speed, flashback, efficiency, NO_x and CO emissions, flue gas dew point, etc. Resolving gas turbine cogenerative group fuels interchangeability, by developing high performance alternative fuel burning technologies, especially hydrogen, will have a major impact over system and environment efficiency. Thus, the studies conducted on more fuels (H_2, CH_4, C_3H_6, C_6H_6, CH_3OH) revealed that [24]: hydrogen and methyl alcohol have the same higher maximum efficiency than other fuels in the same operating conditions; hydrogen fuel has the lowest specific fuel consumption in comparison with other fuels, followed by methane, propen, benzene, and finally methyl alcohol; in the reheating cycle the increase in thermal efficiency is lower than the increased in the intercooling cycle; hydrogen fuel is ideal promising fuel in the gaseous plant which has greater thermal efficiency and greater improvement in the performance of modified gas turbine power plant occurred with intercooling and heat exchanger rather than simple and reheat cycle.

From the point of view of reusing aviation gas turbines for industrial purposes, the possibilities of using liquid fuels are limited, leading to the development of new technologies on gaseous fuels. Thus, the transition of a TV2-117A gas turbine engine from liquid to a gaseous fuel, in order to benefit from landfill gas energy value, has been conducted in many stages: transition from liquid fuel (kerosene) operation to gaseous fuel (natural gas) operation, thus obtaining TA2 gas turbine engine; the transition of TA2 gas turbine engine from natural gas operation to landfill gas operation, thus obtaining TA2 bio gas turbine engine. In order to obtain the two gas turbine engines (TA2 and TA2 bio), numerical simulations were conducted in CFD environment, constructive modifications and gas turbine test bench experiments in order to validate adopted solutions [15]. In this way, TA2 gas turbine engine was integrated in the structure of afterburning installation test bench facilities, from INCDT COMOTI Bucharest (figure 3).

Figure 3. TA2 gas turbine engine mounted on the stand (left) and afterburning installation (right)

Efficiency and reliability are two major parameters that are taken into account since the beginning of a new gas turbine cogenerative group design. In order to obtain a higher efficiency (in balance with cost and reliability), the design team must reach a balance between burning temperature rise and compression ratio, special material selection and complicated cooling systems, customer's specification, etc. Based on a continuum dialogue with Siemens customers, as far back as the predesigned stage, a 36 MWe SGT-750 gas turbine was elaborated [25]. It can be used both in cogenerative applications as well as driving different equipment, stable operation at partial load but also allows the transition to another fuel (operates on dual fuel). Since 2002, Siemens began focusing attention to reliability, so that they give up the high pressure tambour from the recovery boiler (usually used in order to prevent high thermal tensions, a long period of time is imposed in order to reach a certain temperature). Regarding efficiency, flexibility and emission reduction at gas turbine cogenerative groups, important steps were achieved towards: integrating low NO_x burners, lifecycle was analyzed in order to increase efficiency, maintenance interval was enlarged and the transition from one fuel to another was improved at polifuel groups.

3.2. Methods of increasing performances

The development of gas turbine application in various regions of the globe, encouraged researchers to find new methods of increasing performances and to apply new cooling technologies to compressor inlet air, specific to the location. Actually the advantages of inlet air technology application are represented by power losses prevention, losses that occur when ambient temperature exceeds 15 ^{0}C (ISO design temperature), and fuel usage efficiency. In general, the studies of selecting a new method of cooling air take into account each method (see chapter 1) and compare them with a reference case [8]. Evaporative cooling systems imply lower investments, operating and maintenance costs than refrigeration systems but the increase in gas turbine performances is lower. Evaporative cooling system is based on the transition of an air flow through a water soaked environment, system efficiency depending on the surface area of the water soaked environment exposed to air and exposure time. The system is efficient in low humidity regions. In the case of fogging system, the demineralized water is sprayed in the gas turbine inlet air flow, through high pressure nozzles (70 – 200 bar). Figure 4 shows a fogging system, located downstream of the filtering system, in a gas turbine. The air induced through the filtering system reaches the suction chamber, where the water, sprayed in fine droplets (<30 microns), is evaporated in the air flow and produces a cooling effect. In general, the water injected mass flow required represents 2 % from the inlet air, depending on the ambient parameters (temperature and humidity).

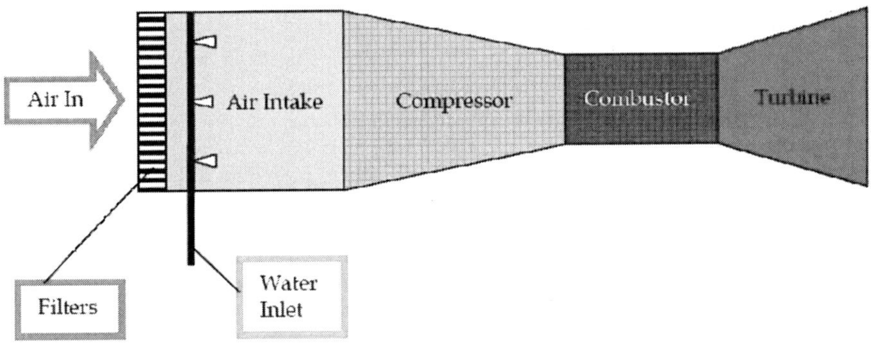

Figure 4. Fogging system scheme, located downstream of the filtering system

Water injection system (with fog) has the advantage of lower pressure drops, higher efficiency, lower refurbishment costs but requires demineralized water, parasite loads occur due to water injection pumps and the control system is complex. Fogging systems, increasingly popular in time, are capable of cooling gas turbine intake air to the temperature of the wet-bulb thermometer, being more efficient than the ones with evaporative media. Although significant investments are required, refrigeration cooling systems (mechanical, absorption, storage) can practically maintain any temperature of gas turbine intake air, recommended in variable humidity regions. In general, fogging system are designed to operate in such a way that the formed droplets should evaporate until entering the compressor, in order to avoid blade damage. In some cases overspray phenomenon can occur (a certain amount of water would not evaporate until entering the gas turbine). In this way overspray phenomenon is similar with water injection between compressor stages (intercooling phenomenon occurring). In certain operating regimes it is possible that evaporative cooling (upstream of the compressor) combined with overspray phenomenon (intercooling) lead to compressor aerodynamic instability (although intercooling effect is added to evaporative cooling) [26]. Combining overspray phenomenon with the usage of a low power heating fuel, steam injection in the combustion chamber or a high level of blade wear can lead to catastrophic results for the gas turbine. Some authors [27] consider that at gas turbine cogenerative cycles, intercooling is recommended, in order to reduce high pressure compressor stage power consumption. Thus, the factors that need to be taken into account when adopting a new technology of cooling the inlet air are [28]: plant configuration (gas turbine engine can be used within an open cycle, a cogenerative cycle, a combined cycle with high operation – intermittent or basic); the amount of compressed air per kW (requiring a small amount of compressed air the cooled air required decreases and, implicitly, exploitation costs decrease), the ambient air enthalpy at given conditions by the design (selecting a high enthalpy of air per kg of dry air leads to a large cooling system, with high exploitation costs); air cooling temperature (for each gas turbine engine model exists an optimum cooling temperatures at given environment conditions).

Although the integration of an afterburning installation (figure 2 and 3) can lead to an increase in cost (with approximately 10-15 % from the cost of the heat recovery steam generator) certain advantages are highlighted: an increase in steam amount at the heat recovery steam generator; thermal energy can be managed more easily; steam amount efficiency, depending on technological process requirements; the heat recovery steam generator can operate even at gas turbine shutdown; it can compensate ambient parameter variations; in some cases can contribute to NO_x emissions reduction by interacting with gas turbine emissions; can burn other fuels usually inadequate to gas turbines (biogas, furnace gas, flammable gases resulted from gasification, etc.).

4. THEORETICAL RESEARCH DEVELOPED AT INCDT COMOTI BUCHAREST

At INCDT COMOTI Bucharest, theoretical research approached issues concerning: processes regarding water spray in gas turbine intake; intercooling; influence of air cooling over gas turbine performances; thermogasodynamic processes from the afterburning installation's burner.

Thus, in order to understand the phenomenon of cooling air at water spraying through an impact body nozzle [29], numerical simulations in CFD environment were conducted, with the working domain consisting in air (gas) and water (liquid). In the first version the impact body (a cone with 1.2 mm base and generator lines) was not taken into account. In the second version the cone was positioned, firstly, at 0.4 mm towards the pipe's end, then moved at 0.8 mm, and 1 mm respectively. Sprayed water reaches in the calculus domain (a cylinder of 1m in diameter and 2.5 m in length) through a 1.2 mm diameter and 12 mm length pipe positioned along the cylinder's symmetry axis. In order to capture as accurately as possible the turbulence phenomenon that occurs in the cone and pipe area, at the same time with water flow sprayed around the cone, the mesh was refined (figure 5).

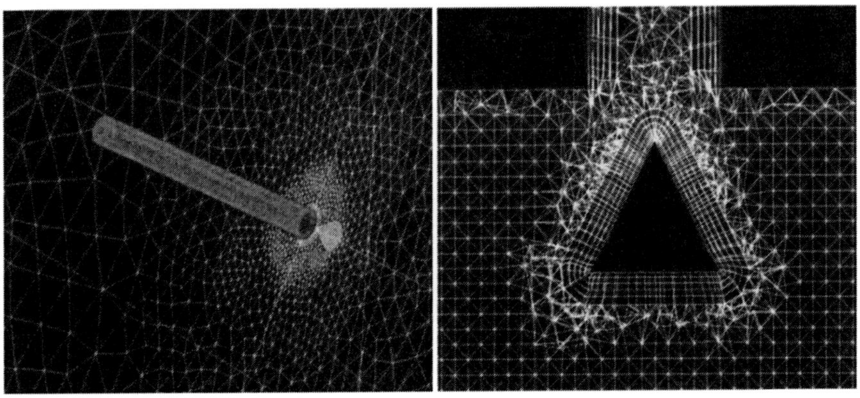

Figure 5. Calculus mesh at the inlet zone of sprayed water in the domain (left) and around the cone (right)

Also, along the pipe and cone walls boundary layers were created, in order for an accurate capturing of the flow near the walls. Water is sprayed into the atmosphere with an axial speed of 15 m/s, as droplets of 100 μm diameter, having a temperature of 288 K. Boundary conditions for the cone and pipe are of smooth adiabatic wall type. The turbulence model used was k-ε. Work methodology was based on the model described in [30]. Numerical simulation results are presented in figures 6 and 7, highlighting the spray without impact body and with impact body positioned at 1 mm distance from the pipe outlet.

Figure 6 reveals that cone insertion leads to pronounced water jet flaring. Cone position modification has a minimum effect over the water spray flare angle. For a better observation of sprayed water jet along the calculus domain, 300 mm in diameter planes were created (comparable with TA2 gas turbine engine intake-figure 3) along the symmetry axis corresponding to z axis coordinate of 0.5 m, 1 m, 1.5 m, 2 m, and 2.5 m respectively.

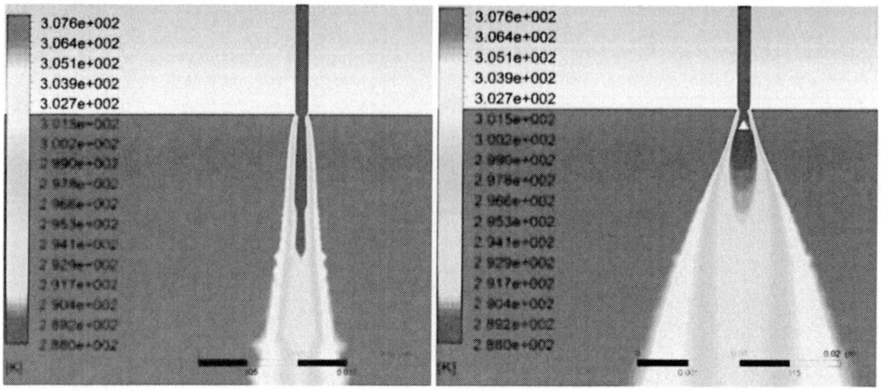

Figure 6. Detail regarding temperature field at water output: without cone (left) and with the cone positioned at 1 mm distance (right)

Figure 7 shows that in case of numerical simulations without the cone, the water jet is directioned along the symmetry axis, while in case of impact body numerical simulations the water jet is more dispersed, as expected. In the case of numerical simulations without the cone, the average temperatures decrease more rapidly.

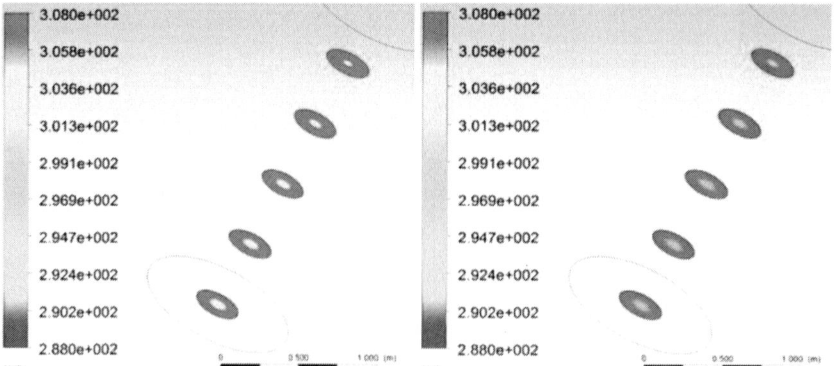

Figure 7. Temperature field in different cross planes: without cone (left) and with the cone positioned at 1 mm distance (right)

Compressor performances are influenced by flow steadiness at the rotor outlet, more specifically if the flow angle varies from hub to shroud than the flow within the diffuser will be unsteady and will lead to separations on the blade vane. Also, the flow at the tip of the blade and at the centrifugal rotor outlet is intensily distorted and unsteady. In order to quantify the impact of intercooling on a two stage compressor, a CFD study was conducted in which the second stage inlet temperature was reduced. Performance impact was monitored, in comparison with the case in which between the two compression stages no cooling was applied. Thus, two CFD analyses were conducted, in which the total inlet second rotor temperature was considered 460 K, for the case without cooling between the compression stages and 313 K, for the case with cooling.. The calculus mesh for this case is a structured mesh and has approximately 1 million elements. The walls are considered to be adiabatic, waterproof and velocity at wall is considered to be zero. In order to improve calculus time, a single channel has been considered (rotor consisting in 15 blades and 15 splitters), using the periodicity function. For spatial discretization was used a second order scheme. A compressible flow was considered in calculus, the governing equations being written as Reynolds Averaged, time and mass averaged. Shear Stress Transport k-ω was considered as turbulence model. SST k-ω model is based on tangential tensions transport. With the help of this turbulence model, accurate results and separation zone dimension (that forms under the influence of high pressure gradients) can be obtained. The results are presented in figures 8-10.

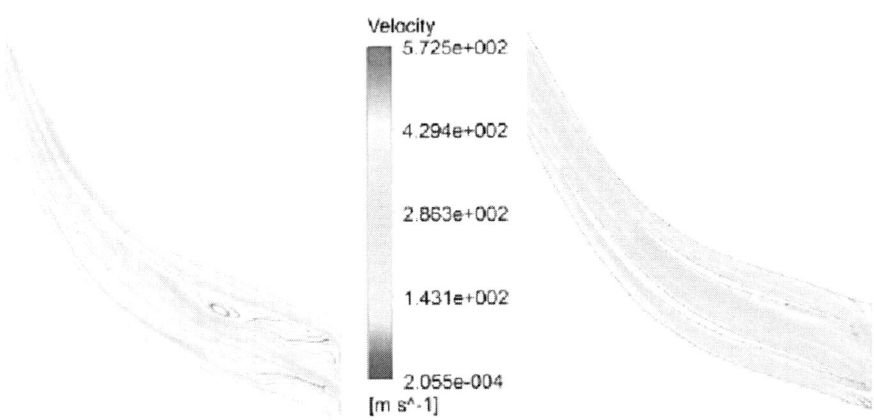

Figure 8. Streamlines, at 50 % of blade height, measured from the hub, in the case of cooling (left) and in the case without cooling (right)

Analyzing the CFD simulations results, it can be observed that along with temperature reduction at rotor inlet, streamlines in the rotor indicate that a recirculation zone is forming on the splitter. This shows that flow angles in the rotor have changed, fact that led to boundary layer separation in the case of intercooling (figure 8). Also, it can be observed (in meridional plane) the fact that Mach number rises at the rotor outlet, from 0.88 in the case without intercooling to 1.055 in the case with intercooling (figure 9).

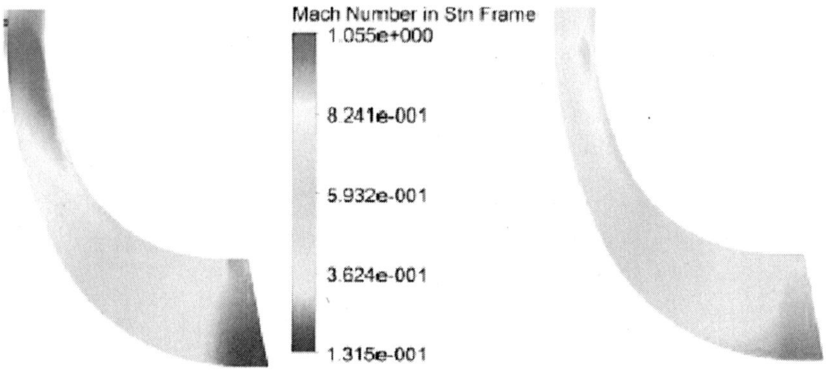

Figure 9. Mach number in meridional plane, in the cooling case (left) and in the case without cooling (right)

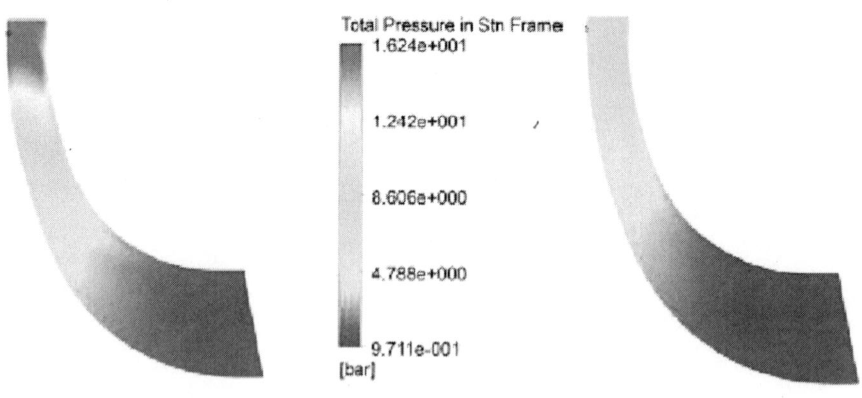

Figure 10. Total pressure in meridional plane, in the cooling case (left) and in the case without cooling (right)

Another difference can be observed also in outlet rotor total pressure, the pressure increases in the case of intercooling (figure 10). Thus, at centrifugal rotor outlet, in the case of intercooling, an increase of 0.2 in Mach number can be observed, reaching 0.938. This indicates a transonic regime at rotor outlet. Corroborated with the fact that the flow angle at rotor outlet reached 12.72^0 from 17.713^0, it means that the existing stator must be redesign.

In general, the performances of a turboshaft engine is compared with the ones given by ideal cycle calculus, thermodynamic analysis including thus, the ideal gas turbine work range. Gas temperature at turbine inlet, as gas turbine outlet section and outlet pressure, are equal in the case of ideal and real gas turbine [31].

Thus, the main configuration to be taken into account for thermodynamic analysis is the one for a monorotor turboshaft engine, with two high and low pressure compressor stages, coupled with a turbine that includes both gas generator stages, as well as power shaft supply stages. For performance calculus, the above configuration is considered (figure 11-left) and intercooling configuration (between compressor low and high pressure stages – figure 11 right). The heat exchanger, that provides intercooling, decreases air temperature in the high pressure compressor inlet section at 40 °C (313 K). Intercooling application decreases the mechanical work consumed by the compressor, without affecting the mechanical work produced by the turbine, leading to an increase in the mechanical load available at the output shaft [32]. In thermodynamic analysis, for the two gas turbine engine configurations presented above, a series of parameters were imposed for operational purposes. The calculus method used is in accordance with [33]. Common conditions to both configurations refer to: ambient temperature, whose variation influences gas turbine performances (shaft power, thermal efficiency, specific fuel consumption); gas temperature in the turbine inlet section, having the same value in all the cases presented, does not vary with ambient temperature, parameter imposed by turbine alloy properties; turbine outlet pressure, in order to ensure exhaust, in the context of pressure losses on the exhaust; burned gases pressure at the exhaust outlet so that the gases can pass through the heat recovery steam generator; coefficients of pressure, speed, energy losses with compressor and turbine efficiencies, and compression ratio (in initial configuration), low pressure compressor (in intercooling configuration) respectively are considered constant. In addition, for the configration that includes intercooling, the following parameters are imposed: temperature at high compressor inlet, the same for all calculus cases, that allows the use of a low temperature resistant material and allows to obtain a high compression ratio at the same mechanical work consumed on the second compressor; the pressure loss coefficient in the heat exchanger; air pressure at combustion chamber inlet, that derives from total compression ratio imposition. In these conditions, the gas turbine output power variation depending on the inlet air temperature (for the various compression ratio), for simple configuration and intercooling and for different compression ratios, is presented in figure 12.

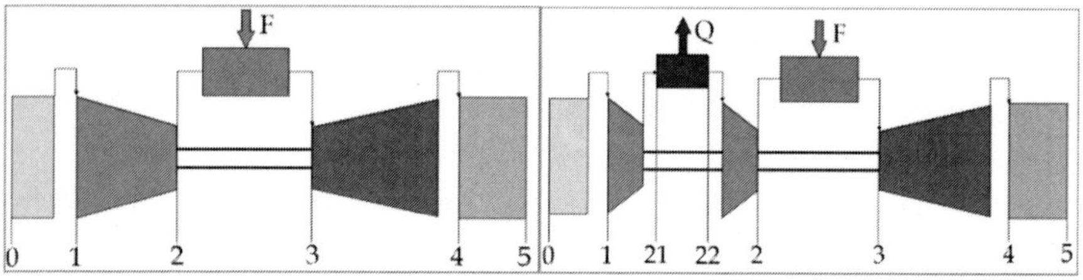

Figure 11. Gas turbine engine configurations – general schemes (0 – intake inlet; 1 – compressor inlet (low pressure compressor); 21 – heat exchanger inlet; 22 – high pressure compressor inlet; 2 – combustion chamber inlet; 3 – turbine inlet; 4 – exhaust inlet; 5 – exhaust outlet)

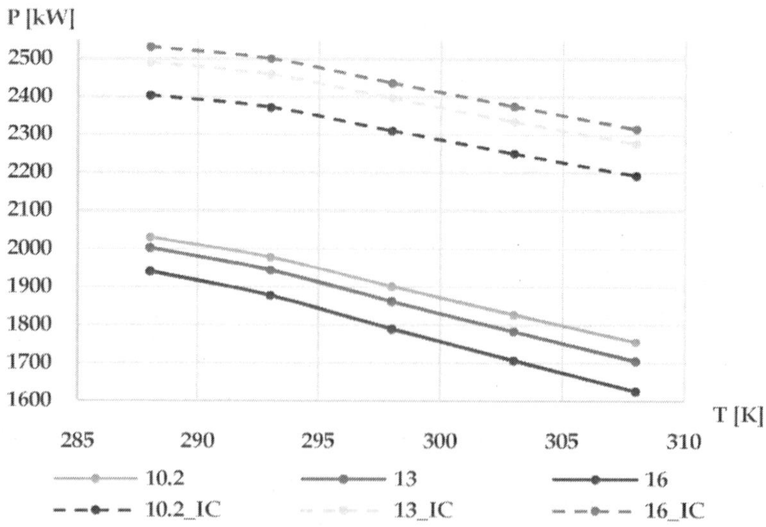

Figure 12. Gas turbine output power variation versus inlet air temperature (at the various compression ratio), for simple configuration and intercooling-IC

It is noticed a decrease of shaft power and thermal efficiency, simultaneous with an increase in specific fuel consumption, at an ambient temperature increase, phenomenon that occurs for all cases in which thermodynamic analysis was conducted, both in the initial gas turbine configuration, as well as in intercooling. For the initial configuration, a pronounced variation occurs with ambient temperature increase, easily noticeable in the case of high compression ratios. In the case of intercooling gas turbine, performance variation is approximately linear for all the three compression ratios, with a decrease in power of 2.5 % for each 5 degrees of the environment.

In the field of afterburning, the theoretical research has soughed: to obtain a reduced pollutants afterburning installation, in comparison with the existing cogeneration power plant 2xST 18 (figure 3); afterburning installation flexibility when supplied with gaseous fuels; to study the influence of water spraying into the gas turbine's combustion chamber upon the afterburning [14, 15]. Thus, the numerical simulations performed in CFD environment have showed that (at nominal conditions), by modifying the afterburning module of the cogenerative plant 2xST 18, the NO_x concentration in the exhaust gases is lowered three times [14, 15]. The base module of the cogenerative plant has been mainly modifying by flaring at 15° and by introducing a concentrator.

5. INLET AIR COOLING INSTALLATION AND AFTERBURNING INTEGRATION WITHIN GAS TURBINE COGENERATIVE GROUP — FUTURE RESEARCH

Gas turbines usually function with high air excess, between 3 and 6 at nominal regime, and even higher at partial-load regimes [34]. This enables the use of an afterburning module, but the control of the air/fuel mixing requires the minimization of the air excess. Thus, reducing the exhaust emissions must be balanced against providing the cogenerative group's performances in a flexible manner.

5.1. Efficiency-emissions binom

The main pollutant emissions produced by gas turbines consist of: nitrogen oxides (NO_x), carbon monoxide (CO), volatile organic compounds (VOCs). Gas turbines typically operate at high loads which make their design for optimum combustion and maximum efficiency to be made at nominal load. Controlling the concentration of all pollutants is difficult in the conditions of variable loads operating. In high loads operating regimes the concentration of NO_x is higher, while in lower loads operating regimes (under 50 %) the thermic efficiency decreases and the concentrations of CO and volatile organic compounds increase. Thus the factors which determine the formation of pollutants in the exhaust gases are [14, 15]: the temperature and the excess of air in the primary zone; the process homogenization degree in the primary zone; the combustion process products residence time; the "freezing" characteristic of the reaction near the fire tube; etc. To decrease the emissions of NO_x is necessary to reduce the temperature in the area in which the combustion reaction takes place and in the high temperature zones, respectively to rethink the distribution of the air flows (combustion in stages). The final version of the gas turbine's combustion chamber will be a compromise between the level of the pollutant emissions, performance and flexibility. Depending on the combustion temperature, in figures 13 and 14 are represented the NO_x and CO emissions levels typical for a class of industrial gas turbines, using different fuels and at various operating regimes [35]. The high CO emissions level indicates an incomplete combustion and a decrease in efficiency. From figures 13 and 14 it can be observed that for temperatures up to 760 ^{0}C, the levels of NO_x and CO are comparable (especially for natural gases), while in the case of higher temperatures the NO_x emissions increases rapidly, while the CO emissions level remains practicly constant. By comparison, in the case of micro gas turbines, operating at 70-100 % loads, the CO emissions are low (figure 15) but they increase fast when operating at under 70 % load [36]. In the case of micro gas turbines the NO_x emissions level are low over a wide range of operating regimes (30-100 % load). Adding a heat recovery steam generator and a cooling system for the intake air (with fog) could be a solution for increasing the performance and decreasing the influence of high temperature in the summer. The high content of vapours in the combustion gases (by injection of water/steam) leads to: acid corrosion (when using fuels which contain sulfur); increase thermal solicitations of the combustion chamber, reduce the heat recovery level, etc. The exhaust gases flow at the gas turbine exit is turbulent and uneven in the transversal section. Thus there might appear reverse flows in some areas of the heat recovery steam generator transveral section. The unevenness of the flow at the combustion chamber exit and the variation of the exhaust gases compozition influence the functioning of the afterburning module. Thus the afterburning is influenced in terms of efficiency, pollutants, flame stability but also in terms of corossion of the elements subject to the action of the exhaust gases. Generally, for a good design of the exhaust gases flow into the heat recovery steam generator, the following factors must be taken into acount [15]: the geometry of the gas turbine exhaust and its direction; the size of the heat exchange surfaces; the position of the afterburning; masic flow and mean speed at the gas turbine exit; local speeds near the walls and at the entrance of the first heat exchange surface. In general the gas turbine's evacuation is not directly coupled with the heat recovery steam generator; after leaving the evacuation of the gas turbine (in the case of the cogenerative power plant 2xST 18 – Suplacu de Barcau, figure 2), the exhaust gases pass through a silencer, a by-pass assembly, an adaption section to the afterburning and then they reach the afterburning chamber. An uniform distribution of the flow in the transversal section insures a proper functioning of the heat recovery steam generator, especcially of its overheater. This creates the necessary premises to ensure low emissions for the cogenerative group. If the exhaust gases coming from the turbine or the air flow are not evenly distributed, in the same way as the fuel, serious variations of the temperature can appear downstream of the burner.

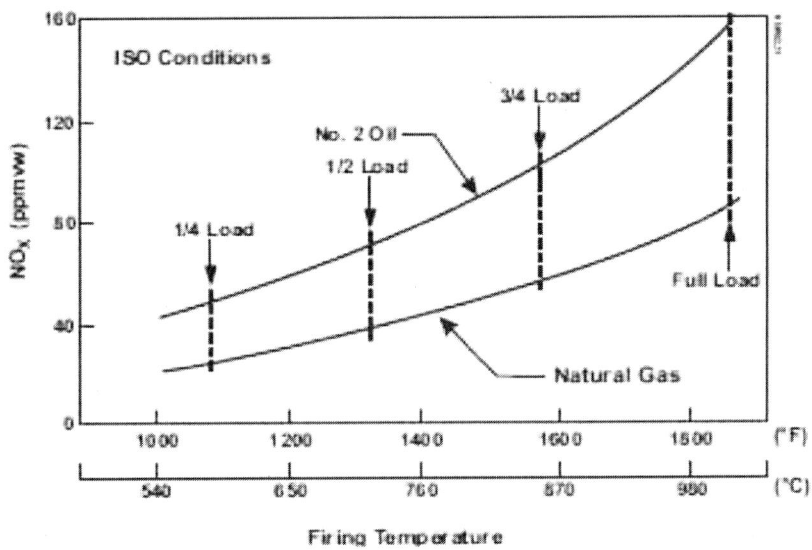

Figure 13. NO_x emissions variation for a class of industrial gas turbines - © 2010 Richard 'Layi Fagbenle. Originally published in [35] under CC BY-NC-SA 3.0 license. Available from: http://dx.doi.org/10.5772/10206

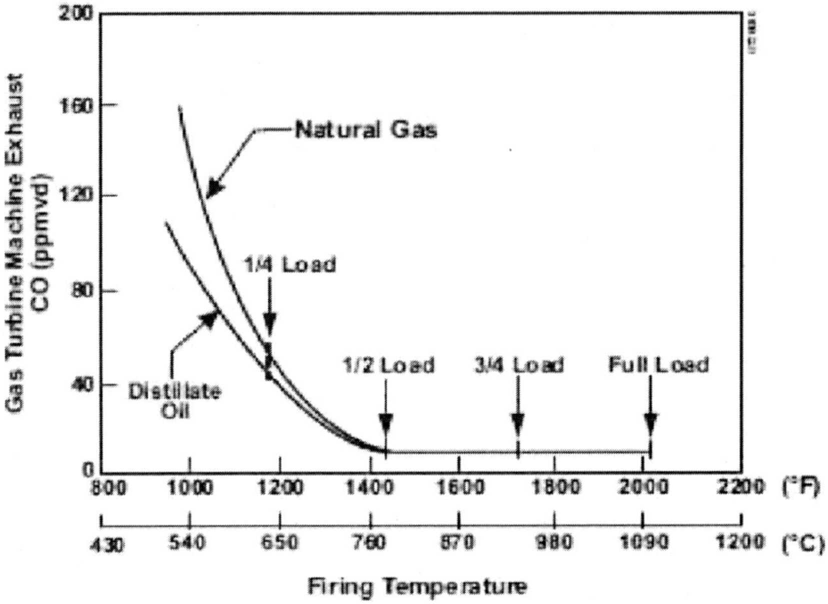

Figure 14. CO emissions variation for a class of industrial gas turbines - © 2010 Richard 'Layi Fagbenle. Originally published in [35] under CC BY-NC-SA 3.0 license. Available from: http://dx.doi.org/10.5772/10206

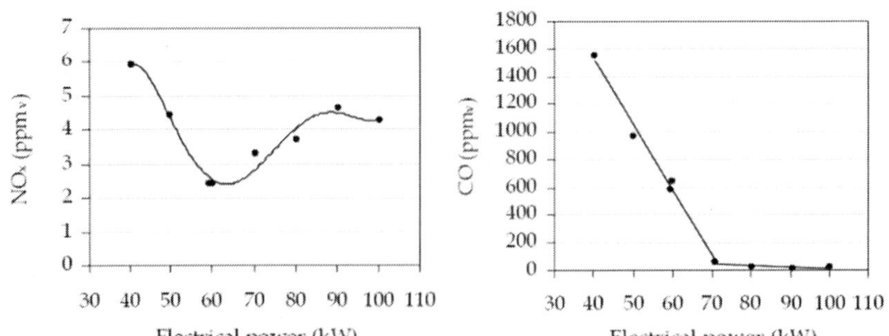

Figure 15. NO_x (left) and CO (right) emissions variation for a class of micro turbines - © 2010 Flavio Caresana, Gabriele Comodi, Leonardo Pelagalli, Sandro Vagni. Originally published in [36] under CC BY-NC-SA 3.0 license. Available from: http://dx.doi.org/10.5772/10211

In general, the variation of the speed (upstream of the burner), on 90 % of the burner section, mustn't exceed ± 15 % of the mean speed on the whole transversal section. In reality the exhaust gases temperature, downstream of the burner, will never be perfectly uniform. Even with a perfect distribution of the gas flow in the turbine, upstream of the burner, the temperature of the gases in the zone of each afterburning module will be higher than the temperature in the areas between the modules. These requirements must fall in the market trends, where (in terms of fluctuation of the electricity and fuel prices) the flexibility in functioning has become a major subject, new concepts being imposed. Increased flexibility aims [37]: rapid start and stop; rapid change of the load; increase reliability in the case of quick start and load predictability; frequency control and of auxiliary services.

5.2. Experimental research conducted at INCDT COMOTI Bucharest

The experimental research carried out at INCDT COMOTI Bucharest, in close relation with the theoretical research (see chapter 4), have concentrated on: obtaining an afterburning installation on gaseous fuel, with low emissions; the realization of water spraying systems and their testing; experimenting water injection in the intake device of the gas turbine, at nominal functioning regime.

The numerical simulations carried out in CFD environment show that, by modifying the afterburning module of the cogenerative power plant 2xST 18, three times lower NO_x emissions can be obtained at nominal functioning regime [14, 15]. Up until now there have been carried out comparative experiments (figure 16) using gaseous fuel, at partial load (3 % of the nominal load). Thus the experiments at partial load show a reduction of the NO_x emissions with 30 %. From figure 16, right, it can be observed that the flame better fills the fire tube of the new afterburning module, and the temperature field is more uniform, being in good correlation with the results.

Before beginning the experiments regarding the effects of water injection in the intake device of the TA2 gas turbine, there has been studied the form of the water spray jet using a liquid fuel atomizer from TV2-117A gas turbine (figure 17). In these tests there has been chosen the atomizer circuit which ensures a droplet diameter (figure 17 left) comparable with the dimension of the gas turbine's intake device (figure 18 left). During the tests the water (untreated) has been sprayed at a pressure of 30±0.5 bar, the atomizer being position on the TA2 gas turbine axis. The distance between the atomizer and the gas turbine intake has been varied between 1500 mm and 2000 mm. The testing rig is composed by: TA2 gas turbine, natural gas afterburning installation – positioned on the vertical, water spraying instalation, command and data acquisition chamber. The emissions measurement has been realized using the Horiba PG 250 gas analyzer, positioned on the stack (the exit of the afterburning chamber – figure 18 right). The experiments conducted on the TA2 gas turbine, using gaseous fuel, have been carried out at starting regime (10500±6 rpm; 75±5 Nm^3/h natural gas flow).

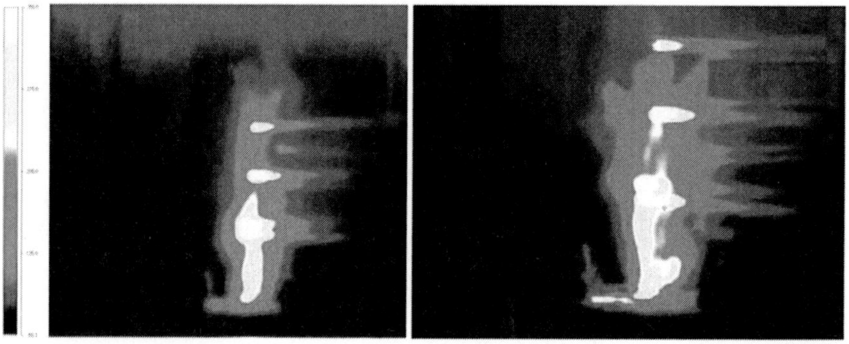

Figure 16. Comparative analysis, in infrared, of the afterburning modules (the existing module on the 2xST power plant – left; numerical simulation result – right) [15, 38]

Figure 17. Water spraying tests, using a liquid fuel atomizer from TV2-117A gas turbine

The mean temperature T_{3M} (the gases mean temperature before the gas turbine) has been determined using 17 thermocouples (figure 19 left), using the methodology presented in [15]. The free turbine speed has rezulted from the vibration analysis, data acquisition being realized unsing module IEPE BNC Ni9233 (National Instruments). The exhaust gases temperature (T_5 – figure 19 right) has been measured using a thermocouple.

Figure 18. Experiments regarding the injection of water in TA2 gas turbine (left) and gas analyser PG 250 (right)

The experiments included several series of 20 minutes in which the gas turbine functioned at starting regime (without the afterburning module), using natural gases as fuel (for 10 minutes the gas turbine functioned without water injection, followed by 10 minutes of functioning with water injection). The results regarding NO_x emissions during the experiments conducted on the TA2 gas turbine using natural gases as fuel, with/without water injection (the atomizer is positioned on the gas turbine's axis, at 1500 mm from the gas turbine's intake device) are presented in figure 20. The results obtained so far regarding water injection in the intake device of TA2 gas turbine (for power increase) are more qualitative, although by introducing water the speed of the free turbine is increased with about 20 rpm. Simultaneously with the introduction of water, temperatures T_{3M} and T_5 decrease with about 5 °C, and the NO_x emissions decrease with about 4 ppm. It is expected that the influences will be more conclusive when the load of the gas turbine and the water quantity are increased.

Figure 19. The thermocouples position for determining T_{3M} (left) and T_5 (exhaust device exit - right)

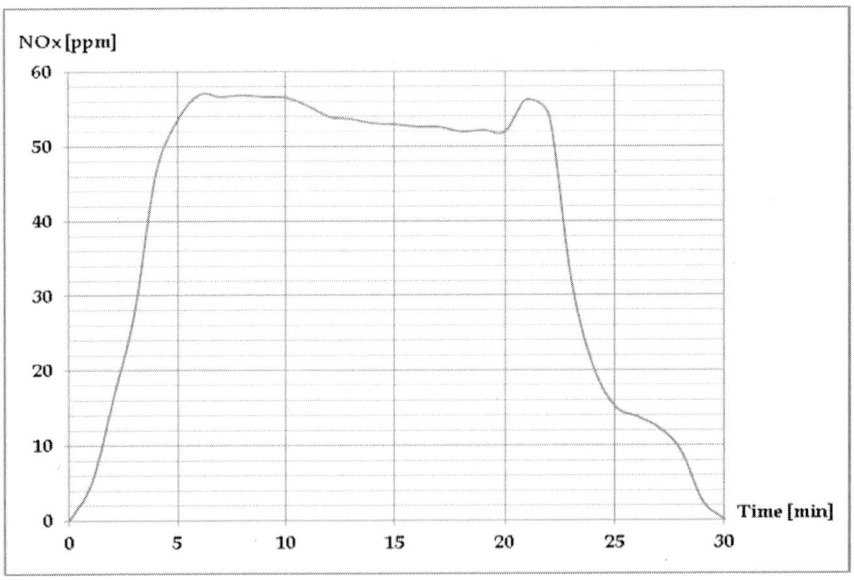

Figure 20. NO_x emissions variation during the experiments carried out on TA2 gas turbine, with water injection in the intake device

After about 10 series of experiments, of 20 minutes each, the boroscopic examination of the gas turbine didn't revile any significant deposits. Possibly in the future it will be necessary to appropriately treat the water injected in the intake device.

5.3. Future research

Future research to be conducted at INCDT COMOTI Bucharest will follow the general context given by the efficiency-emissions reduction-flexibility triad, by numerical simulations and experiments concerning: processes that take place during the cooling of the intake air; combustion in the gas turbine and afterburning installation; increase of the efficiency and pollutants reduction; flexibility regarding the used fuels.

6. CONCLUSIONS

1. The selection of a location for a gas turbine cogenerative plant imposes climatic conditions and demands adequate technical solutions to meet performance requirements, especially during summer season when inlet air temperature rises, leading to a decrease in power output and efficiency.

2. Water content modifies thermodynamic properties of intake air (density, specific heat) affecting power output and heat mass flow resulted from the gas turbine. If in the past air humidity was neglected, in present day cogenerative gas turbine power increase, but also water/steam injection impose the need for it to be taken into account.

3. The main research directions in the area of cogenerative groups with gas turbines efficiency increase are: combustion temperature increase; compression ratio increase; improvement of design methods, combustion technology and advanced materials; technological transfer for aviation domain to industrial turbines domain; integrated systems (combined cycles, intake air cooling, exhaust turbine gases heat recovery, afterburning, etc.).

4. Determinant factors concerning the overall efficiency of the cogenerative group are: gas turbine exit temperature, temperature at the heat recovery steam generator stack; ambient environment temperature. For these the most influential factor upon the

increase of the overall efficiency is the temperature at the heat recovery steam generator stack.

5. Operating flexibility of equipment has become a major subject. Gas turbines are designed to function generally at nominal regime, in maximum efficiency conditions and minimum pollutants. At cogenerative groups with heat recovery steam generator, for producing technological steam, is preferable that the flexibility to the process demands to be achieved using afterburning installation.

6. Theoretical and experimental research conducted at INCDT COMOTI Bucharest, allowed: to be showed that, in the case of a gas turbine with intercooling, the performances variation is approximately linear for a compression ratio between 10.2-16, with a power decrease of 2.5 % for each 5 degrees increase of the environment temperature; to be obtained a afterburning module with a 30 % reduction of the NO_x reduction (at partial load) in comparison with the existing cogenerative power plant 2xST 18 – Suplacu de Barcau; to demonstrate the power increase and NO_x emissions reduction when injecting water in the intake device of TA2 gas turbine.

7. ACKNOWLEDGEMENTS

The research have been made based on contracts 22-097/2008, 22-108/2008 (Programme "Partnerships in priority fields") and 7N/2009, financed by Romanian Ministry of Education, Research, Youth and Sports. The consortium involved in contracts 22-097/2008, 22-108/2008 includes several companies in Romania: INCDT COMOTI Bucharest, SC OMV PETROM SA, UPB-CCT Bucharest, SC OVM-ICCPET SA Bucharest, SC ICEMENERG SA Bucharest, SC ERG SRL Cluj, SC TERMOCAD SRL Cluj.

REFERENCES

1. Technology characterization: Gas turbines, 2008, http://www.epa.gov/chp/documents/catalog_chptech_gas_turbines.pdf (accessed 17 July 2014).
2. Notiuni introductive privind ciclurile termodinamice ale motoarelor termice si turbinelor cu gaze. Diagrame entropice de stare, determinarea randamentelor termodinamice, http://www.rasfoiesc.com/inginerie/tehnica-mecanica/Notiuni-introductive-privind-c67.php (accessed 17 July 2014).
3. Contemporary Cogeneration Technologies, http://www.energymanagertraining.com/Journal/02112007/ContemporaryCogenerationTechnologies.pdf (accessed 25 June 2014).
4. Abam F., Ugot I., Igbong D., Performance analysis and components irreversibilities of a (25 MW) gas turbine power plant modeled with a spray cooler, American J. of Engineering and Applied Sciences 2012; 5 (1): 35-41, http://thescipub.com/html/10.3844/ajeassp.2012.35.41 (accessed 17 June 2014).
5. Pankaj P., Better power generation from gas turbine along with improved heat rate, International Conference–PowergenIntl 2000-Florida, Bangkok, ASME 2003-Atlanta, http://www.energymanagertraining.com/announcements/issue25/winners_papers_Issue25/05_PankajKPatel.pdf (accessed 17 June 2014).
6. Hosseini R, Beshkani A., Soltani M., Performance improvement of gas turbines of Fars (Iran) combined cycle power plant by intake air cooling using a media evaporative cooler, Energy Conversion and Management 2007; 48: 1055–1064,
7. Bagabir A., Khamaj J., Hassan A., Experimental and theoretical study of micro gas turbine performance augmentation, Emirates Journal for Engineering Research 2011; 16 (2): 79-88, http://www.engg.uaeu.ac.ae/ejer/issues/V16/pdf_iss2_16/7%20Ahmed%20Bagabir-22-22-F11_EXPERIMENTAL_Formatted%20_2__authors.pdf (accessed 22 June 2014).
8. P. dos Santos, Andrade Cl., Zaparoli E., Comparison of different gas turbine inlet air cooling methods, World Academy of Science, Engineering and Technology 2012; 6, http://www.waset.org/publications/2686/comparison-of-different-gas-turbine-inlet-air-cooling-methods.pdf (accessed 17 June 2014).

9. Etminan V., Moghiman M., Bajestan E., Boghrati M., Performance improvement of simple cycle gas turbine by using fogging system as intake air cooling system, http://www.google.ro/url?sa=t&rct=j&q=&esrc=s&source=web&cd=95&ved=0CGMQFjAEOFo&url=http%3A%2F%2Fprofdoc.um.ac.ir%2Farticles%2Fa%2F102366.doc&ei=lTpaU7KUFeOI7AbsjYHQDQ&usg=AFQjCNHCYaE5k7lhoIXTYF2D03ETtWw-ww (accessed 17 June 2014).

10. Espanani1 R., Ebrahimi S., Ziaeimoghadam H., Efficiency Improvement Methods of Gas Turbine, Energy and Environmental Engineering 2013; 1(2): 36-54, 2013 DOI: 10.13189/eee.2013.010202, http://www.hrpub.org/download/201309/eee.2013.010202.pdf (accessed 17 June 2014).

11. Gopinath V., Navaneethakrishnan G., Performance Evaluation of Gas Turbine by Reducing the Inlet Air Temperature, International Journal of Technology Enhancements and Emerging Engineering Research 2013; 1 (1): 20-24, http://www.ijteee.org/final-print/aug2013/Performance-Evaluation-Of-Gas-Turbine-By-Reducing-The-Inlet-Air-Temperature.pdf (accessed 20 June 2014).

12. Al-Doori W., Parametric Performance of Gas Turbine Power Plant with Effect Intercooler, Modern Applied Science 2011, 5 (3): 173-184

13. Kyprianidis K., Future Aero Engine Design: An Evolving Vision, Advances in Gas Turbine Technology, Dr. Ernesto Benini (Ed.), InTech, 2011, ISBN: 978-953-307-611-9, DOI: 10.5772/19994. Available from: http://cdn.intechopen.com/pdfs-wm/22893.pdf (accessed 27 June 2014).

14. Barbu E., Vilag V., Popescu J., Ionescu S., Ionescu A., Petcu R., Cuciumita C., Cretu M., Vilcu C., Prisecaru T, Afterburning Installation Integration into a Cogeneration Power Plant with Gas Turbine by Numerical and Experimental Analysis, Advances in Gas Turbine Technology, Dr. Ernesto Benini (Ed.), InTech, 2011, ISBN: 978-953-307-611-9, DOI: 10.5772/19994. Available from: http://www.intechopen.com/books/advances-in-gas-turbine-technology/afterburning-installation-integration-into-a-cogeneration-power-plant-with-gas-turbine-by-numerical- (accessed 17 June 2014).

15. Barbu E., Petcu R., Vilag V., Silivestru V., Prisecaru T., Popescu J., Cuciumita C., Tomescu S., Gas Turbine Cogeneration Groups Flexibility to Classical and Alternative Gaseous Fuels Combustion, Progress in Gas Turbine Performance, Dr. Ernesto Benini (Ed.),

InTech, 2013, ISBN: 978-953-51-1166-5, Available from: http://www.intechopen.com/books/progress-in-gas-turbine-performance/gas-turbine-cogeneration-groups-flexibility-to-classical-and-alternative-gaseous-fuels-combustion (accessed 17 June 2014).

16. Motahar S., Alemrajabi A., Performance augmentation of an aero-engine gas turbine using steam injection, International Conference on Energy and Environment 2006 (ICEE 2006), http://www .iasj.net/iasj?func=fulltext&aId=57646 (accessed 27 June 2014).

17. Tyagi K., Khan M., Effect of gas turbine exhaust temperature, stack temperature and ambient temperature on overall efficiency of combine cycle power plant, International Journal of Engineering and Technology 2010; 2 (6): 427-429, http://www.enggjournals .com/ijet/docs/IJET10-02-06-18.pdf (accessed 17 June 2014).

18. Begovic M., Maintaining declared performance in gas turbines during increased ambient temperatures, Energija 2009; 58 (2): 192-207.

19. Kurz R., Gas turbine performance, Procedings of the thirthy-fourth turbomachinery Symposium, 2005, 131-146, http://turbolab.tamu. edu/proc/turboproc/T34/t34-14.pdf (accessed 17 June 2014).

20. Pavri R., Moore G., Gas turbine emissions and control, http://site.ge-energy.com/prod_serv/products/tech_docs /en/downloads/ger4211.pdf (accessed 27 June 2014).

21. Homji C., Mee Th., Gas turbine power augmentation by fogging of inlet air, http://turbolab.tamu.edu/proc/turboproc/ T28/Vol28010.pdf (accessed 27 June 2014).

22. Kumar P., Krishna A., Raju G., Comparative Analysis on Performance of a Gas Turbine Power Plant with a Spray Cooler, International Journal of scientific research and management (IJSRM) 2013; 1 (7): 354-358.

23. Ionel I., Ungureanu C., Popescu F. Analiza nivelului de emisii poluante prin schimbarea combustibilui la cuptoarele de tratament termic. http://www.tehnicainstalatiilor.ro/articole /images/nr12_ 76-82. pdf (accessed June 14, 2014).

24. 24. Al-dawoodi M., Theoretical study for the effect of different fuels on the perfomance of open gas turbine power plant, http://www.google.ro/url?sa=t&rct=j&q=&esrc=s&source=web&c d=119&ved=0ciybebywcdhu&url=http%3a%2f%2fwww.uobabylo

n.edu.iq%2fpublications%2fapplied_edition4%2fpaper_ed4_32.doc &ei=3unau5halir07abm_yh4aq&usg=afqjcnho5qxani118jnp9p733f wu7ki5uq (accessed 17 June 2014).

25. Anderson L., Gas turbine design for cogeneration balancing efficiency and reliability for maximum economy, http://www .gasnet.com.br/conteudo/11938/Gas-turbine-design-for-cogeneration-balancing-efficiency-and-reliability-for-maximum-economy (accessed 27 June 2014).

26. Brun K., Kurz R., Gas turbine life limiting rffects on inlet and interstage water injection, Procedings of the thirthy-fourth turbomachinery Symposium 2005, 45-52, http://turbolab.tamu.edu/proc/turboproc/T34/t34-06.pdf (accessed 27 June 2014).

27. Ibrahim Th., Rahman M., Alla A., Study on the effective parameter of gas turbine model with intercooled compression process, Scientific Research and Essays 2010; 5(23): 3760-3770, http://www. academicjournals.org/article/article1380620575_Ibrahim%20et%20al.pdf (accessed 27 June 2014).

28. Dharap A., Ghanwat A., Inlet Air cooling for Gas turbines, Air Conditioning and Refrigeration Journal, October-December 1999, http://www.ishrae.in/journals/1999oct/article01.html (accessed 27 June 2014).

29. Suryan1 A. et all., Experimental investigations on impaction pin nozzles for inlet fogging system, Journal of Mechanical Science and Technology 2011; 25 (4): 839-845.

30. Makky A., Spray modeling tutorial using ANSYS-CFX, 2012, http://www2.warwick.ac.uk/fac/sci/eng/study/pg/students/esrhaw/spray_modelling_using_ansys.pdf (accessed 27 June 2014).

31. Wilson C.D., Riggins D.W. Performance characterization of turboshaft engines: work potential and second-law analysis. American Helicopter Society 58th Annual Forum. Montreal, Canada, June 11-13. 2002.

32. Boyce M. P., Gas Turbine Engineering Handbook, Fourth Ed., Butterworth-Heinemann, 2012, ISBN 978-0-12-383842-1.

33. Pimsner V., Maşini cu palete (Bladed machines). Ed. Tehnică, Bucharest, 1988.

34. Carlanescu C., Manea I., Ion C., Sterie St. (1998). Turbomotoare – Fenomenologia producerii si controlul noxelor, Editura Academiei Tehnice Militare, Bucuresti

35. Fagbenle R., (2010). Exergy and Environmental Considerations in Gas Turbine Technology and Applications, Gas Turbines, Gurrappa Injeti (Ed.), ISBN: 978-953-307-146-6, InTech, DOI: 10.5772/10206. Available from: http://www.intechopen.com /books/gas-turbines/exergy-and-environmental-considerations-in-gas-turbine-technology-and-applications (accessed 27 June 2014).

36. Caresana F, et all, (2010). Micro Gas Turbines, Gas Turbines, Gurrappa Injeti (Ed.), ISBN: 978-953-307-146-6, InTech, DOI: 10.5772/10211. Available from: http://www.intechopen .com/books/gas-turbines/micro-gas-turbines-mgts-(accessed 27 June 2014).

37. Henkel N, Schmid E., Gobrecht E., Operational flexibility enhancements of combined cycle power plants, 2007, https://www.cee.siemens.com/web/ua/ru/production_energy/Documents/Service_Operational%20Flexibility%20Enhancements%20of%20Combined%20Cycle%20Power%20Plants.pdf (accessed 27 June 2014).

38. Barbu E., Fetea G., Petcu R., Vilag V., Dragasanu L., Afterburning installation of 2xST 18 cogeneration power plant – investigations on combustion and NOx emissions, Chemical Engineering Transactions 2013; 34: 37-42,

8

CHAPTER

ANALYSIS OF GAS TURBINE BLADE VIBRATION DUE TO RANDOM EXCITATION

E.A. Ogbonnaya[1], R. Poku[1], H.U. Ugwu[2], K.T. Johnson[3], J.C. Orji[3] and N. Samson[3]

[1] Department of Marine/Mechanical Engineering, Niger Delta University, Wilberforce Island, Bayelsa State, Nigeria

[2] Department of Mechanical Engineering, Michael Okpara University of Agriculture, Umudike, Abia State, Nigeria

[3] Department of Marine Engineering, Rivers State University of Science and Technology, Port Harcourt, Rivers State, Nigeria

1. INTRODUCTION

In recent times, a considerable impact has been made on the modeling of dynamic characteristics of rotating structures. Some of the dynamic characteristics of interest are critical speed, systems stability and response to unbalance excitation. In the case of Gas Turbines (GT), the successful operation of the engine depends largely on the structural integrity of its rotor shaft (Surial and Kaushal, 2008).

The structural integrity in turn depends upon the ability to predict the dynamic behavior or characteristic accurately and meet the design requirement to withstand steady and vibratory stresses. An accurate and reliable analysis of the rotor shaft behavior is therefore essential and requires complex and sophisticated modeling of the engine spools rotating at different speeds, static structure like casing, frames and elastic connections simulating bearing (Zhu and Andres, 2007). In this work, GT rotor shaft dynamic modeling will be based on the speed and the force response due to unbalance. During the design stage of GT rotor shaft, the dynamic model is used to ensure that any potential harmful resources are outside the engine operating speed.

Engine vibration tests are part of the more comprehensive engine test program conducted on all development and production engines (Surial and Kaushal, 2008). In the design and retrofit process, it is frequently desirable and often necessary to adjust some system parameters in order to obtain a more favourable design or to meet the new operating requirement Kris, et al (2010). Rotor shaft unbalance is the most common reason in machine vibration (Ogbonnaya 2004).

Most of the rotating machinery problem can be solved by using the rotor balancing misalignment. Mass unbalance in a rotating system often produces excessive synchronous forces that reduce the life span of various mechanical elements (Hariliaran and Srinivasan, 2010). A very small amount of unbalance may cause severe problem in high speed rotating machines. Overhung rotors are used in many engines ring applications like pumps, fans, propellers and turbo machinery. Hence, the need to consider these problems, even at design stages.

The vibration signature of the overhung rotor is totally different from the center hung rotors. The vibration caused by unbalance may destroy critical parts of the machine, such as bearings, seals, gears and couplings. In practice, rotor shaft can never be perfectly balanced because of manufacturing errors, such as porosity in casting, and non-uniform density of materials during operation (Eshleman and Eubanks (2007), Mitchell and Melleu (2005), Lee and Ha (2003)).

1.1. Damped unbalance response analysis

The second part in the rotor shaft dynamic analysis is conducting the damped unbalance response analysis. The objective of this analysis is to accruably determine the critical speeds and the vibration response (amplitude and phase angle) below the trip speed. API 617 (2002) requires that damped unbalance response analysis be conducted for each critical speed within the speed range of 0%-125% of trip speed. The standard requires calculating the amplification factors using the half power method described in figure 1. This helps to determine the required separation margin between the critical speed and the running speed.

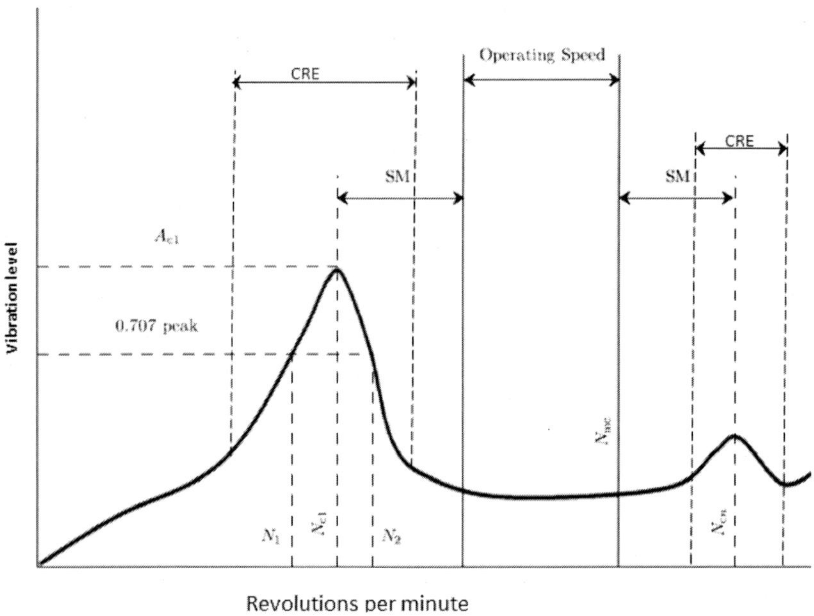

Figure 1. Amplification factor calculation from API 617 (2002)

The Legends in figure 1 are as follows:

N_{C1}=Rotor first critical, center frequency, cycles per minute

N_{Cs}=Critical speed,

N_{mc}=Maximum continuous speed, 105%

N_1=Initial (lesser) speed at 0.707 x peak amplitude (critical)

N_2=Final (greater) speed at 0.707 x peak amplitude (critical)

N_2-N_1=Peak width at the ball-power point

A_f=Amplification factor

=*N C*1 *N* 2 - N 1 =NC1N2 - N1

SM =Separation margin

CRE =Critical response envelope

A_{c1}=Amplification at N_{c1}.

A_{c2}=Amplification at N_{c2}

As the amplification factor increases, the required speed increase up to a certain limit. A high amplification factor (AF>10) indicates that the rotor vibration during operation near a critical speed could be considerable and that critical clearance component may rub stationary elements during periods of high vibration.

From the figure 1, a low amplification factor (AF<5) indicates that the system is not sensitive to unbalance when operating in the vicinity of the associated critical speed. To ensure that a high amplification factor will not result in rubbing the standard requires that the predicted major axis peak to peak unbalance response at any speed from zero to trip speed does not exceed 75% of the minimum design diametric running clearances through the compressor.

This calculation is be performed for different bearing clearance and lubricating oil temperatures to determine the effect of the rotor stiffness and damping variation on the rotor shaft response. Also, the standard requires an unbalance response verification test for rotor shaft operating above the critical speeds.

The test results are used to verify the accuracy of the damped unbalance response analysis in terms of the critical speed location and the major axis amplitude of peak response. The actual critical speeds shall not deviate by more than 5% from the predicted, as the actual vibration amplitude shall not be higher than the predicted value (Bader, 2010).

The purpose of this study is therefore to show the dynamic response of a GT rotor shaft using a mathematical model. In the course of this work, it was noted that the rotor shaft can never be perfectly balanced because of manufacture errors. Hence the model involved the following:

a. The working principle of GT rotor shaft

b. Causes of unbalancing on a rotor shaft

c. The response of the system to the critical speed

The objectives and contributions to learning through this work are also as follows:

a. To model the dynamic response of a GT turbine system

b. To consider defects of the rotor shaft on the components of the GT system.

c. The result of this research could thereafter be extended to solve problems on other rotor dynamic engines.

d. To propound a viable proactive integrated and computerized vibration-based maintenance technique that could prevent sudden catastrophic failures in GT engines from rotor shaft.

2. ROTOR SHAFT SYSTEM AND UNBALANCE RESPONSE

The rotor shaft system of modern rotating machines constitutes a complex dynamic system. The challenging nature of rotor dynamic problem has attracted many scientists, Engineers to investigations that have contributed to the impressive progress in the study of rotating systems.

According to Ogbonnaya (2004), the study of the unbalance responses of GT rotor shaft is of paramount importance in rotor dynamics. He further stated that the GT rotor shaft is a continuous structure and cannot therefore be considered as an idealized lumped parameter beam. Hariliarau and Srinivasan (2010) gave detailed model of rotor shaft coupling. They reviewed the rotor shaft and coupling modeled using Professional Engineer wildfire with the exact dimension as used in experimental setup. A number of analytical methods have been applied to unbalance response such as the transfer matrix method, the finite element method (Lee and Ha, 2003) and the component model synthesis method (Rao, et al 2007; Ogbonnaya, et al 2010).

Unbalance response investigations of geared rotor bearing systems, based on the finite element modeling was carried out by Neriya, et al (2009) and Kahraman, et al (2009) utilizing the model analysis technique. Besides, based on the transfer matrix modeling, Lida et al (2009) and Iwatsubo et al (2009) reported on studies utilizing the usual procedure of solving simultaneous equations while Choi and Mau, (2009) utilized the frequency branching technique to carry out the same analysis.

Further concerning unbalance response investigations of dual shaft rotor-bearing system coupled by bearing, (Hibner, 2007 and Gupta et al., 2003) carried out investigation utilizing the usual procedure of solving simultaneous equations based on transfer matrix modeling. However, all of the above investigations resulted in full numerical solutions of the unbalance response of coupled two shaft rotor bearing systems. On the other hand, Rao (2006) suggested analytical closed-form expressions for the major and minor axis radii of the unbalance response or bit for one-shaft rotor bearing.

2.1. Active balancing and vibration control of rotor system

It is well established that the vibration of rotating machinery can be reduced by introducing passive devices into the system (Gupta, *et al,* 2003). Although an active control system is usually more complicated than a passive vibration control scheme, an active vibration control technique has many age advantages over a passive vibration control technique.

First, active vibration control is more effective than passive vibration control in general (Shiyu, and Jianjun 2001). Second, the passive vibration control is of limited use if several vibration modes are excited. Finally, because the active actuation device can be adjusted according to the vibration characteristics during the operation, the active vibration technique is much more flexible than passive vibration control.

2.1.1. Active balancing techniques

A rough classification of the various balancing methods is shown in figure 2. The most recent development in active balancing is summarized in the dashed-lines shown in figure 2.

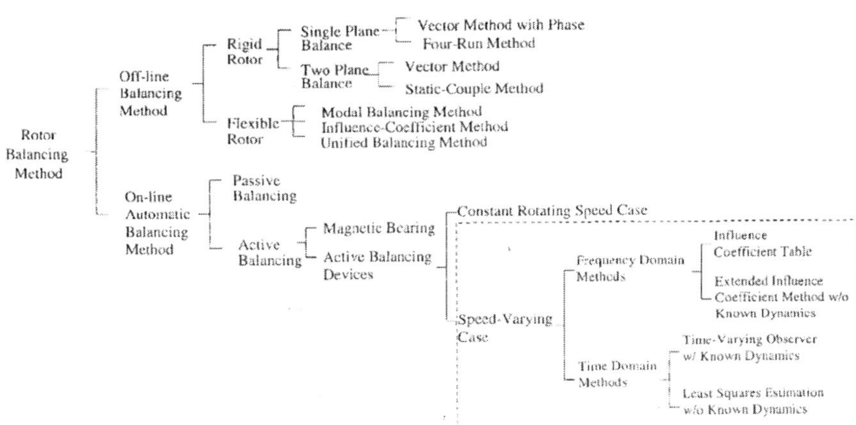

Figure 2. Classification of balancing methods; *Source: Shiyu and Jianjun* (2001)

The rotor balancing techniques can be classified as offline balancing methods and real-time active balancing methods. Since active balancing methods are extensions of off-line balancing methods, a review of off-line methods thus is provided (Shiyu and Jianjun, 2001).

2.1.2. Off-line balancing methods

The off-line rigid rotor balancing method is very common in industrial applications. In this method, the rotor is modeled as a rigid shaft that cannot have elastic deformation during operation. Theoretically, any imbalance distribution in a rigid rotor can be balanced in two different planes. Methods for rigid rotors are easy to implement but can only be applied to low-speed rotors, where the rigid rotor assumption is valid. A simple rule of thumb is that rotors operating under 5000 rpm can be considered rigid rotors. It is well known that rigid rotor balancing methods cannot be applied to flexible rotor balancing. Therefore, researchers developed modal balancing and influence coefficient methods to off-line balance flexible rotors.

Modal balancing procedures are characterized by the use of the modal nature of the rotor response. In this method, each mode is balanced with a set of masses specifically selected so as not to disturb previously balanced, lower modes. There are two important assumptions: (1) the damping of the rotor system is so small that it can be neglected and (2) the mode shapes are planar and orthogonal. The first balancing technique similar to modal balancing was proposed by Hibner (2007). This method was refined in both theoretical and practical aspects in Ogbonnaya (2004).

Many other researchers also published works on the modal balancing method, including Rao (2006). Their work resolved many problems with the modal balancing method such as how to balance the rotor system when the resonant mode is not separated enough, how to balance the rotor system with residual bow, how to deal with the residual vibration of higher modes, and how to deal with the gravity sag. An excellent review of this method can be found in Rao (2006). Most applications of modal balancing use analytical procedures for selecting correction masses. Therefore, an accurate dynamic model of the rotor system is required. Generally, it is difficult to extend the modal balancing method to automatic balancing algorithms.

2.2. Self-excitation and stability analysis

The forces acting on a rotor shaft system are usually external to it and independent of the motion. However, there are systems for which the exciting force is a function of the motion parameters of the system, such as displacement, velocity, or acceleration (Ogbonnaya, 2004). Such systems are called self-excited vibrating systems since the motion itself produces the exciting force. The instability of rotating shafts, the flutter of turbine blades, the flow induced vibration of pipes and aerodynamically induced motion of bridges are typical examples of the self-excited vibration (Rao, 2006).

2.3. Dynamic stability analysis

A system is dynamically stable if the motion or displacement coverage or remains steady with time. On the other hand, if the amplitude of displacement increases continuously (diverges) with time, it is said to be dynamically unstable (Ogbonnaya, 2004). The motion diverges and the system becomes unstable if energy is fed into the system through self-excitation. To see the circumstances that lead to instability, we consider the equation of motion of a single degree of freedom system as shown in equation 1:

$$M\ddot{x} + C\ddot{x} + kx = 0 \tag{1}$$

If solution of the form *x*(+)*C st e* x+Cest when C is a constant, assuming the equation 1 lead to characteristic equation

$$S^2 + \frac{C}{m}S + \frac{k}{m} = 0 \tag{2}$$

The root of this equation is as shown in equation 3:

$$S_{12} = \frac{C}{2m} + \frac{1}{2}\left[\left(\frac{C}{m}\right)^2 - 4\left(\frac{K}{m}\right)\right]^{1/2} \tag{3}$$

Since the solution is assumed to be *x*(+)*C st e* x+Cest , the motion will be diverging and a periodic, if the roots S_1 and S_2 are complex conjugates with positive real parts. Analyzing the situation, let the roots S_1 and S_2 of equation 2 be expressed as:

$$S_1 = P + iq, \quad S_2 = P + iq \tag{4}$$

Where p and q are real numbers so that:

$$(S-S_1)(S-S_2)=S^2-(S_1S-S_2S+S_1S_2)=m+\frac{k}{m}$$
$$=S^2+\frac{C}{m}S+\frac{k}{m}=0 \tag{5}$$

Equations 4 and 5 therefore become

$$\frac{C}{m}=(S_1+S_2)=-2P_1\frac{k}{m}=S_1S_2=P^2+q^2 \tag{6}$$

From equation (6), it is shown that for negative *P* 1 ,*c m* / P1,cm must be positive and for positive *P* 2 +*q* 2 , *km* P2+q2, km , must be positive. Thus the system will be dynamically stable if C and k are positive (assuming that M is positive).

Figure 3. Picture of the used balancing plane; *Source: Shiyu and Jianjun* (2001)

2.3.1. Balancing operation and result

The necessary mass was added at the chosen shaft end shown in figure 3 in order to determine the desired dynamic behaviour. Due to the relatively small rotor radius, it was necessary to use a significant mass; otherwise the obtained influence would have been too low.

2.4. Rotor dynamic model of the shaft line

The complete shaft line was modeled (figure 4) by using the MADYN 2000 software. The model was based on scaled drawings. The four fluid film bearings were calculated with the ALP3T program. The static load of the different bearings was determined by aligning the shaft only, taking into account the flexibility of the different rotors, in a way that the couplings are free of bending moments. The present oil film thickness at nominal speed was not considered in this static calculation.

All bearing pedestals were modeled as pure stiffness and mass. This assumption was tested by performing impact tests on the bearing structure in both vertical and horizontal direction. No resonance frequencies were detected below 50Hz or at multiples of this frequency. Therefore it was decided to determine the static stiffness obtained at 50Hz and use this value for the complete frequency range of the different calculations.

There was only poor rotor dynamic information given by the manufacturer, so it was not possible to completely verify the model. However, the calculated results fitted both the basic available rotor dynamic info arid the measured vibration data quite well. From the calculated eigen values at 50Hz two modes seemed to be present near the operating speed of the shaft line (figure 4).

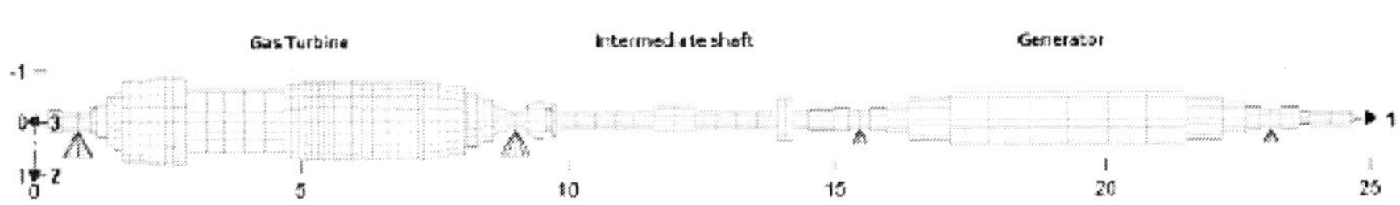

Figure 4. Rotor dynamic shaft line model; *Source: Kris, et. al* (2010)

The closest modes influencing the dynamic behaviour are at respectively 51.9Hz and 53.3Hz. The first eigen mode is the second vertical bending mode of the gas turbine and is unlikely to be causing high vibrations near the generator shaft end. The second eigen mode is a horizontal bending mode of the shaft end. From the eigenvalue analysis it is clear that the damping factor is poor and the mode deformation is almost completely planar (whirling factor of 0). This mode shape is shown in figure 5 and shows clearly the planar deformation near the shaft end. As shown in figure 6, shaft lines can also be represented in 3-D mode shape.

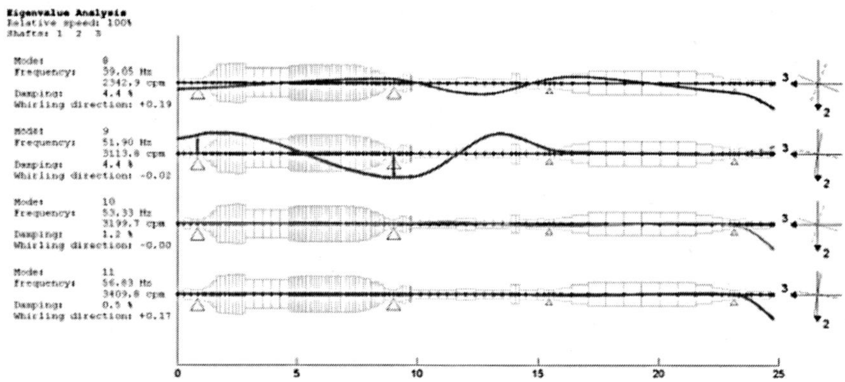

Figure 5. Eigenvalue analysis of the shaft line; *Source: Kris, et. al* (2010)

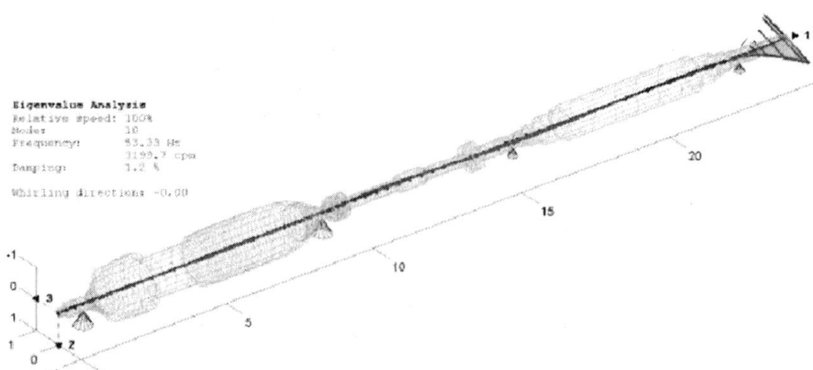

Figure 6. 3D mode shape of the bending critical of the shaft end; *Source: Kris, et. al* (2010)

Based on the measured direct orbit shapes, it was possible to conclude that it was this shaft end mode that was responsible for the high shaft vibrations causing the automatic shutdown of the unit during startup (figure 7). One can clearly see that there is a local horizontal deformation at the generator nondestructive examination measuring plane near nominal speed; but for all the other measuring planes, the relative shaft vibration amplitudes remain rather low.

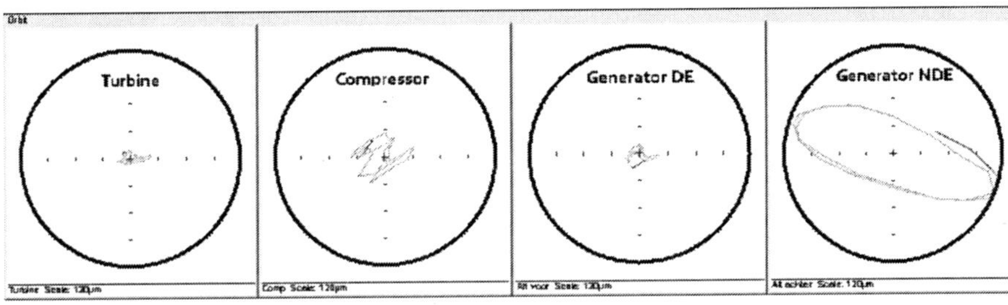

Figure 7. Direct orbit plots at 2980rpm; *Source: Gupta, et al* (2003)

From these results, it was also clear that the "usual" balancing planes at both ends of the generator rotor were not recommended to balance this present unbalance. However, from local inspections of the shaft end, it became clear that it would be possible to do a balancing test run with a mass connected at the end of the shaft line on a rather non-conformistic plane (figure 8). This plane would be the best location because the mode deformation is the highest at the shaft end. This location would also make the balancing plane easily accessible for further adjustment of the mass.

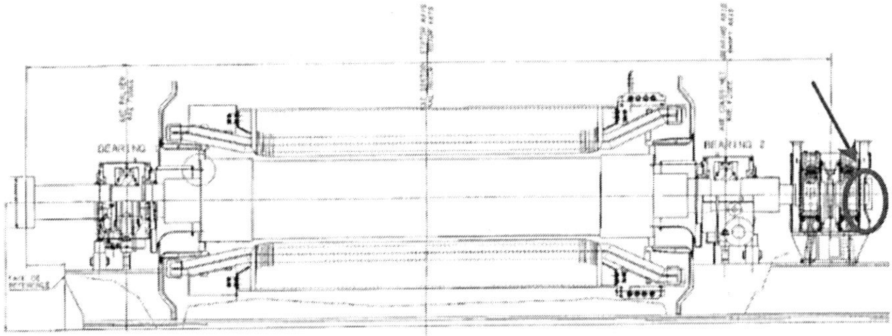

Figure 8. Chosen balancing plane of the generator shaft end; *Source: Gupta, et al* (2003)

An unbalance calculation was done in order to have an idea of the expected response of the generator. The used weight for this calculation was determined by applying a GT unbalance; based on the weight of this free shaft end (figure 9).

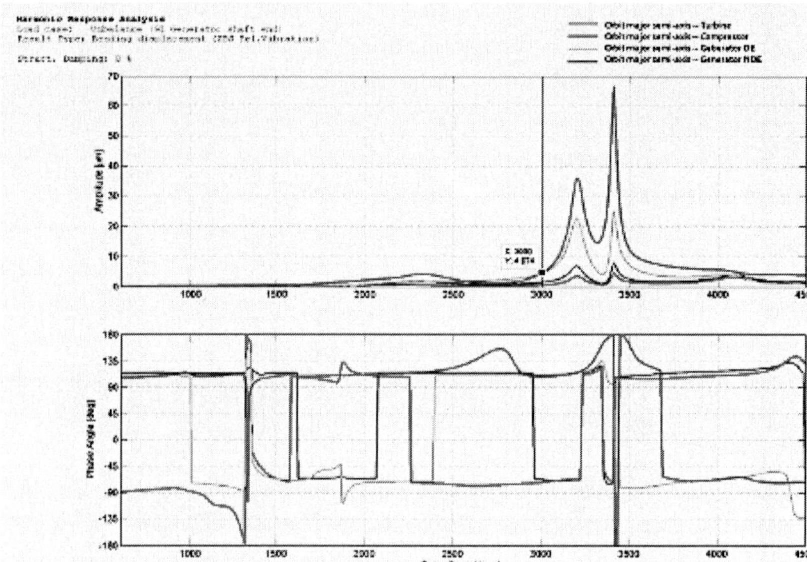

Figure 9. G1 unbalance of the shaft end; *Source: Gupta, et al.* (2003)

This unbalance response shows that there is a certain influence at 50Hz due to this added unbalance, however to balance the shaft end, a correction mass of about 15 times higher would be needed to bring down the actual shaft vibration amplitudes to acceptable levels. The resulting dynamic behaviour of the shaft line is shown in figure 10 and shows that there is an important reduction of the relative shaft vibrations around 3000rpm. This confirms of course that this shaft end mode was the main reason for the increased shaft vibrations.

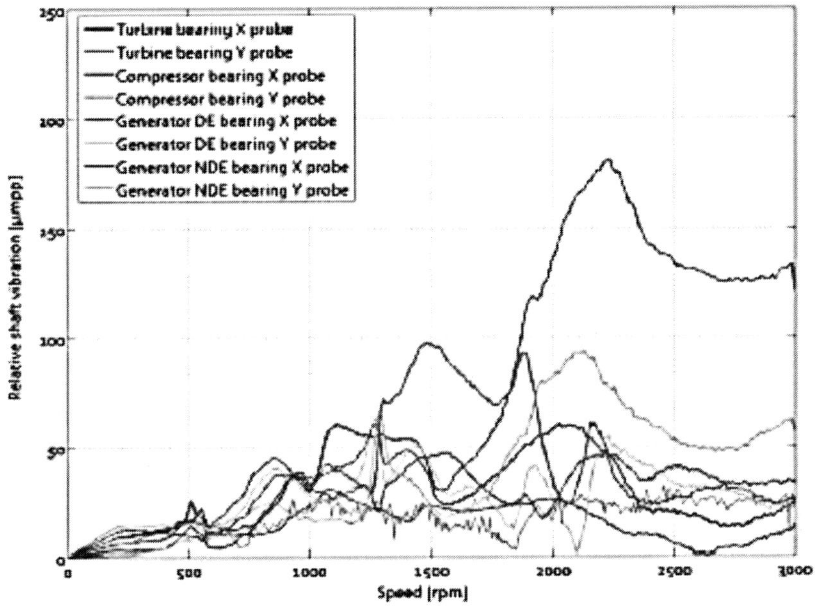

Figure 10. Relative shaft vibration amplitudes after balancing; *Source: Gupta, et al.* (2003)

This balancing correction was done completely remotely. Only a local maintenance responsible went on-site to attach the balancing weight. This made it possible to react quickly on this vibration issue and reduced the unforeseen downtime of the unit to a minimum. The unit could be restarted the same day without any vibrations alarms and a more extensive balancing of the present residual unbalance of the rotor could be scheduled at a more appropriate moment in the maintenance planning.

3. METHODOLOGY FOR BLADE VIBRATION ANALYSIS

Rotor shafts are amongst rotor dynamic components subjected to perhaps the most arduous working condition in high performance rotating equipment used in process and utility plants such as high-speed compressor, steam and gas turbines generators, pumps, etc. Although usually quite robust and well designed, shafts in operation are sometimes susceptible to serious defects that develop without much warning.

In exploitation of rotating machines, some of the observed phenomena are considered to be particularly undesired from the view point of effectiveness and safety. Excessive stress concentrations and rubbing effects occurring between stator and rotors attached to flexible shafts subjected to lateral vibrations can be given as examples of such a detrimental behavior. The modern responsible and heavily affected rotating machines must assure possibly high level of reliability, durability and safety in operation. For these reasons, their design process should be performed very thoroughly in order to obtain relatively small magnitude of unavoidable dynamic excitation, e.g. due to residual unbalance, gas pressure force or electromagnetic force. While aiming at realistic modeling of rotor shaft systems, the actual stochastic nature of important model parameters should be taken into account. In the previous section, different references have been made to the problems of rotor shaft dynamic modeling. A lot of reasons were given as regards to the need to the modeling approach; however, all-inclusive approach must be employed to tame the problems. This involves the understanding of the theoretical model of a rotor shaft, assessment of rotor-shaft vibration due to uncertain residual unbalances and as well as the modeling dynamic response analysis. Therefore, all the modification discussed in this chapter only affects the GT rotor shaft system.

3.1. Theoretical model of a bowed rotor shaft

A typical rotor with a bowed unbalanced shaft is presented in Ogbonnaya (2004). All phase angles are measured with respect to a reference timing mark on the shaft. Suppose the shaft has a residual bow of ∂r ∂r and a phase of ϕ_r, then the mass centre of the disk would be displaced by a distance, e_u from the shaft centre line. This results in a dynamic unbalance response as the shaft rotates.

Suppose the magnitude and phase angle of the combined electrical and mechanical run-out respectively, then the total observed response is:

$$\partial = \partial r + \partial s + \partial o \quad \partial = \partial \mathrm{r} + \partial \mathrm{s} + \partial \mathrm{o} \tag{7}$$

Where:

∂r=bow vector

∂s=response to the mass unbalance vector and the bow vector

∂o=run-out vector

Assumptions:

1. No gyroscopic pull occur (since the disc always rotates in its own plane)
2. Shaft mass is considered negligible (compared to the rigid disc mass)
3. Both shaft and disc rotate with uniform angular velocity, ω
4. The supports are taken as rigid

The equation of motion for a simple rotor with imbalance and shaft bow has been used to obtain the steady state non-dimensionalised rotor response as a function of rotor speed as follows:

$$\bar{z} = \frac{\bar{\delta}_r e^{-i\varnothing r} + f^2 e^{-i\varnothing m}}{1 - f^2 + 2i\xi f} \tag{8}$$

Where:

ϕ_o=angle between the rotor run-out vector and the timing mark (equation 9)

ϕ_r=angle between the bow vector and the timing mark

ϕ_m=angle between the mass unbalance vector and the timing mark

ϕ= phase angle between the shaft centerline response vector and the timing mark (equation 9)

f= frequency ratio,

$\omega\ \omega$ cr / = rotational speedrotor critical speed ωωcr = rotational speedrot or critical speed

ξ ξ

=damping ratio,

$c\ c$ cr / cccr

$\partial\ r$ ∂r

=residual bow

C=shaft damping, Nm/rad

C_{cr} Ccr

=critical speed (with low compensation C)

Consider a shaft with electrical and/or mechanical run-out. The run-out vector is also non-dimensionalized by an unbalance eccentricity given by:

$\delta o - = \delta o e u$ δo-=δoeu

Where

$e u$ eu

=unbalance eccentricity vector (this run-out vector is constant and independent of the shaft rotational speed).

Using the principles of superposition, the constant response due to run-out may be added to equation 3 for steady state response to yield:

$$\bar{z} = \frac{\bar{\delta}_r e^{-i\varnothing r} + f^{2} e^{-i\varnothing m}}{1 - f^{2} + 2i\xi f} + \bar{\delta}_o e^{-i\varnothing o} \qquad (9)$$

If $\gamma = \phi_r - \phi_m$, then equation 9 becomes:

$$\bar{z} = \alpha_r \bar{\delta}_r e^{-i\varnothing r} + \alpha_r e^{-i\varnothing m} + \alpha_r \delta_o e^{-i\varnothing o} \qquad ()$$

Where:

αr αr

= influence coefficient due to bow $= \frac{1}{(1 - f^{2} 2i \xi f)}$

αm αm =influence coefficient due to mass unbalance=

$f2\ (1-f2\ 2i\xi f)$ f2(1−f2 2iξ f)

$\alpha 0$ α0

=influence coefficient due to run-out=1

Separating

z ⁻ z⁻

in equation 10 into real and imaginary components $\left(\bar{z} = {}_r + \mathrm{i}_{\mathrm{i}}\right)$, yields

$$\bar{z}_r = \frac{\left(\bar{\delta}_r, \cos\varphi_r + f^2 \cos\varphi_m\right)\left(1-f^2\right) - 2\varsigma f\left(\bar{\delta}_r, \sin\varphi_r + f^2 \sin\varphi_m\right)}{\left(1-f^2\right)^2 + \left(2\varsigma f\right)^2} + \bar{\delta}_0 \cos\varphi_0 \tag{11}$$

$$\bar{z}_i = \frac{\left(\bar{\delta}_r, \sin\varphi_r + f^2 \sin\varphi_m\right)\left(1-f^2\right) + 2\varsigma f\left(\bar{\delta}_r, \cos\varphi_r + f^2 \cos\varphi_m\right)}{\left(1-f^2\right)^2 + \left(2\varsigma f\right)^2} + \bar{\delta}_0 \sin\varphi_0 \tag{12}$$

Therefore, the shaft amplification factor and phase angle are respectively:

$$A = \sqrt{\bar{z}_r + \bar{z}_i^2} \tag{13}$$

And

$$\theta = \tan^{-1}\left[\bar{z}_i / \bar{z}_r\right] \tag{14}$$

respectively

For the response of the bowed rotor at slow roll, substitute *f*=0 into equation 15 to get:

$$\bar{Z}_o = \bar{\delta}_r e^{-i\bigcirc r} + \bar{\delta}_o e^{-i\bigcirc m} \tag{15}$$

$$\therefore \qquad \bar{z}_r\Big|_o = \bar{\delta}\cos\phi_r + \bar{\delta}\cos\phi_o \tag{16}$$

$$\text{and} \qquad \bar{z}_i\Big|_o = \bar{\delta}\cos\phi_r + \bar{\delta}\cos\phi_o \tag{17}$$

Note: for the case of bowed shaft, the run-out compensator and subtractor compensate for shaft run-out, thereby presenting the response for the rotor as if it had only unbalance with no run-out. Therefore, equations 9 and 10 are the quantities the compensator subtracts from the total response of a bowed rotor at all other speeds. Let this quantity be designated by

z¯ c z¯c

for a bowed rotor compensated for electrical/mechanical run-out.

$$\therefore \qquad \bar{z}_c = \bar{z} - \bar{z}_0 = \frac{\delta_r e^{-i\varphi_r} + f^2 e^{-i\varphi_m}}{1 - f^2 + 2i\xi f} - \delta_r e^{-i\varphi_r} \tag{18}$$

Rearranging equation (18) and combing like terms yields:

$$\bar{Z}_c = \frac{\left(f^2 - 2i\xi f\right)\bar{\delta}_r e^{-i\varnothing m}}{1 - f^2 + 2i\xi f} \tag{19}$$

which can also be put as:

$$\bar{Z}_c = \alpha_{rc}\bar{\delta}_r e^{-i\varnothing r} + \alpha_m e^{-i\varnothing m} \tag{20}$$

Where

$\alpha\, rc$ αrc

=influence coefficient for compensated bow and given by:

$$\alpha_{rc} = \frac{\left(f^2 - 2i\xi f\right)}{1 - f^2 + 2i\xi f} \text{ as } \alpha_{rc} \rightarrow \alpha_m \text{ for } large\ f. \qquad \text{as}$$

$\alpha\, rc \rightarrow \alpha\, m$ for*largef*. αrc→αmforlargef.

The response due to compensated bow and an equal unbalance eccentricity are equal.

3.2. Mathematical modeling of dynamic response of GT rotor shaft

When a system is subjected to force harmonic excitation, its vibration response takes place at the same frequency as that of the excitation. Common sources of harmonic excitation are: unbalance of the rotating shaft, forces produced by reciprocating machines, or the motion of the machine itself. These vibrations are undesirable to equipment whose operation may be distributed. Resonance is to be avoided in general, and to prevent large amplitudes, vibration isolators are often used.

3.2.1. Response of single degree of freedom system

According to Ogbonnaya (2004), a rotor shaft can be modeled using the single degree of freedom system shown in figure 11(a) and the free body diagram as presented in figure 11(b).

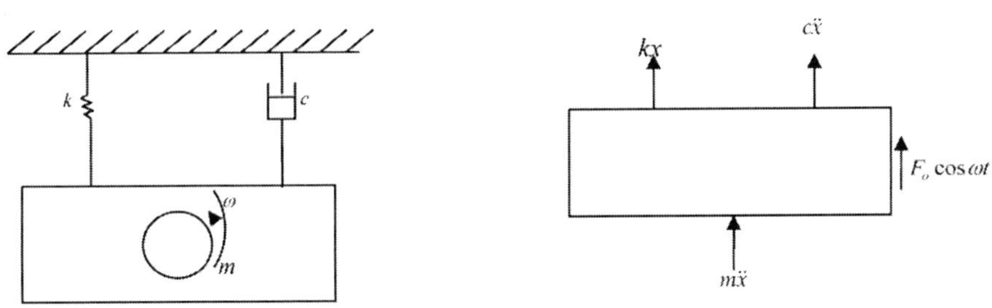

(a) Model of rotor shaft (b) Free body diagram

Figure 11. Model of a rotor shaft system

From the figures above, the equation of motion can be stated as shown in equation 21.

$$m\ddot{x} + c\dot{x} + kx = 0 \tag{21}$$

3.2.2. Impulse response approach

Figures 12 (a) and (b) show the forcing function in the form of a series of impulses and unit impulse excitation at t=ζ, while figure 12 (c) represents the impulse response function.

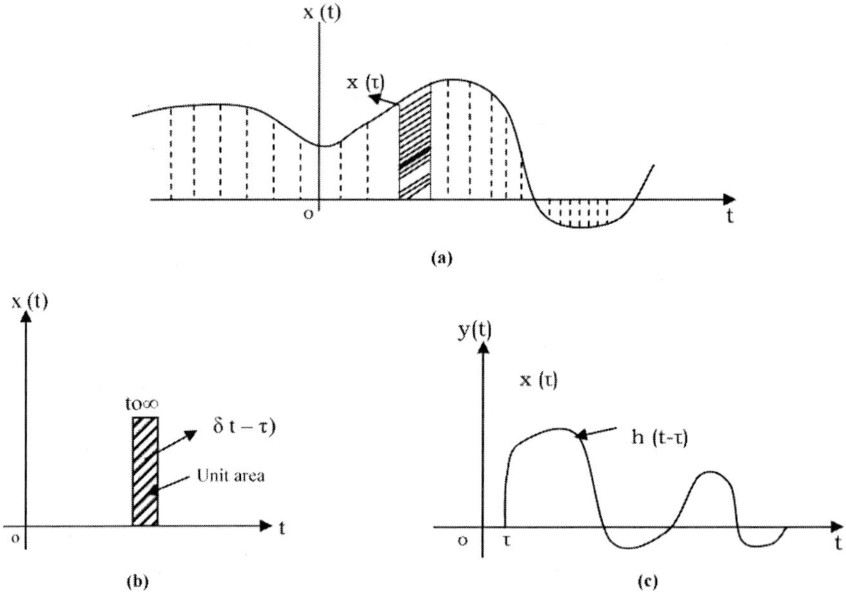

Figure 12. (a) forcing function in the form of series of impulses; (b) Unit impulse excitation at t=τ; (c) Impulse response function; *Source: Sadhu, (2006)*

Here, consideration was given to the forcing function x(t) to be made up of a series of impulse of varying magnitude as shown in figure 12a. Let the impulse applied at time τ be denoted as *x (τ) dτ*. If *y (t)=H (t-τ)* denotes the response of the unit impulse excitation, δ (t-τ), it is called the impulse response function. The response to the total excitation is:

$$y(t)=\int_{-\infty}^{t} x(\tau)h\,(t - \tau)d\tau \qquad (22)$$

Since h (t-τ)=0 when t < τ or τ > t, the upper limit of integration can be replaced by ∞ so that:

$$y(t)=\int_{-\infty}^{\infty} x(\tau)h(t-\tau)d\tau \tag{23}$$

By changing the variables from τ to θ=t – τ, equation 23 can be written as:

$$y(t)=\int_{-\infty}^{\infty} x(t-o)h(o)d\theta \tag{24}$$

The response of the system y(*t*) can be known if the impulse response function h (t) is known.

3.2.3. Frequency response approach

In this case, the transient function x(*t*) can be expressed in terms of its Fourier transform x(*ω*) as:

$$x(t)=\frac{1}{2\pi}\int_{\omega=-\infty}^{\infty} x(\omega)e \tag{25}$$

Consider the forcing function of unit modulus as:

$$\bar{x}(t)=e^{i\omega t} \tag{26}$$

Its response can be denoted as;

$$\bar{y}(t)=H\left(\omega\right)e^{i\omega t} \tag{27}$$

Where: H (ω) is called the complex frequency response function. Thus, the total response is:

$$y(t)=H(\omega)x^{t}=\int_{-\infty}^{\infty} H(\omega)\frac{1}{2\pi}x\left(\omega\right)e^{i\omega t}d\omega \tag{28}$$

If y (ω) denotes the Fourier transform of the response function y (t), then:

$$y(t)=\frac{1}{2\pi}\int_{-\infty}^{\infty} x\left(\omega\right)e^{i\omega t}d\omega \tag{29}$$

Comparing equations 27 and 28 we get:

$$x(\omega)=H(\omega)x(\omega)$$

(30)

Equation (30) can be used to find the response of the system once H (ω) is known.

3.2.4. Computation of dynamic response of GT rotor shaft system

From the model of the rotor shaft system shown in figures 11(a) and (b), the equation of the motion for equilibrium can be written as:

$$\sum F = ma = m\ddot{x}$$

(31)

F o cos*ωt*+*ω*−*k*(*d st* +*x*)=*mx* ¨ Focosωt+ω−k(dst+x)=mx¨

Where *d st* = dst= elongation of the spring (meter)

F o cos*ωt*+*ω*−*kd st* −*kx*=*mx* ¨ Focosωt+ω−kdst−kx=mx¨

But

ω=*kd st* ω=kdst

F o cos*ωt*+*kd st* −*kd st* −*kx*=*mx* ¨ Focosωt+kdst−kdst−kx=mx¨

F o cos*ωt*−*kx*=*mx* ¨ Focosωt−kx=mx¨

$$m\ddot{x} + kx = F_o \cos \omega t$$

(32)

By considering damping force, equation (32) can be further expressed as:

$$m\ddot{x} + c\dot{x} + kx = F_o \cos \omega t$$

(33)

$$\ddot{x} + \frac{c}{m}\dot{x} + \frac{k}{m}x = \frac{\ddot{F}_o \cos \omega t}{m}$$

(34)

but $cm = 2\mu$ cm=2μ and $km = \omega n\ 2$ km=ωn2 (Ogbonnaya, et al; 2013); where μ μ is the slip factor which is the parameter describing how much the rotor exit flow angle deviate from the actual blade angle; i.e. fouling/corrosion factor. Hence, equation (34) is modified as:

$$\ddot{x} + 2\mu\dot{x} + \omega_n^{\ 2} x = 0 \tag{35}$$

Equating the complementary function to zero, gives:

$$x_1 = C_1 e^{\left[-\mu + \sqrt{\mu^2 - \omega^2}\right]t} + C_2 e^{\left[-\mu + \sqrt{\mu^2 - \omega^2}\right]t} \tag{36}$$

$$x_1 = C_1 e^{\left[-\mu + \sqrt{\mu^2 - \omega^2}\right]t} + C_2 e^{\left[-\mu + \sqrt{\mu^2 - \omega^2}\right]t}$$

$$x_2 = [C_1 + C_2 t] e^{-\mu t}$$

$$x_3 = e^{-\mu t}[A\cos\sqrt{(\omega^2 - \mu^2)}t + B\sin\sqrt{(\omega^2 - \mu^2)}t$$

The general solution of equation (34) thus, is given by:

$$X = \frac{F_o}{m}\left[\frac{\sqrt{4\mu^2\omega^2 + (\omega_n^{\ 2} - \omega^2)^2}}{4\mu^2\omega^2 + (\omega_n^{\ 2} - \omega^2)^2}\cos(\omega t - \alpha)\right] \tag{37}$$

The overall solution is a combination of the complementary function and the general solution given as follows:

$$X = \frac{F_o}{m\sqrt{4\mu^2\omega^2 + (\omega_n^{\ 2} - \omega^2)^2}} \tag{38}$$

The vibration displacement amplitude is observed as:

$$F_o = XM\sqrt{4\mu^2\omega^2 + (\omega_n{}^2 - \omega^2)^2} \tag{39}$$

where:

$$F_o = XM\sqrt{4\mu^2\omega^2 + (\omega_n{}^2 - \omega^2)^2} \tag{40}$$

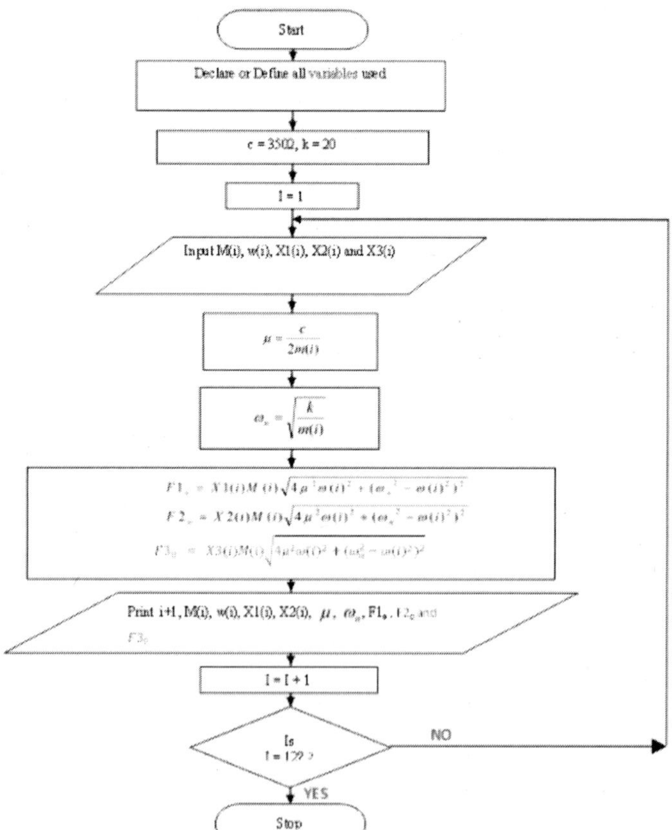

Figure 13. Program Flow Chart of the rotor Shaft Dynamic Response

Table 1. Operational Data from Healthy GT 17 of Afam Power Station

Time (hr)	Speed (RPM)	Active load (KW)	Natural frequency, (ω_n)	Vibration amplitude (Br1) (mm)	Vibration amplitude (Br2) (mm)	Vibration amplitude (Br3) (mm)
01.00	3063	50	321.7	4.8	6.2	0.9
02.00	3076	50	322.3	4.8	6.2	0.8
03.00	3074	50	321.7	4.6	6.2	0.9
04.00	3077	50	322.3	4.9	6.2	0.8
05.00	356	50	320.4	4.9	6.2	0.9
06.00	3001	50	316.7	5.1	6.3	1.0
07.00	3063	45	321.0	5.1	6.5	1.0
08.00	3076	40	322.3	5.3	6.5	0.9
09.00	3028	40	318.6	5.0	6.7	1.2
10.00	3081	36	323.6	5.1	6.4	0.9
11.00	3053	35	319.8	5.2	6.5	1.0
12.00	3.53	45	316.0	5.2	6.4	1.2
13.00	3005	37	314.2	5.2	6.6	1.7
14.00	3060	42	319.8	5.0	6.7	0.9
15.00	3056	41	319.2	5.3	6.5	0.9
16.00	3021	41	319.2	5.1	6.6	0.9
17.00	3077	40	322.9	51.2	6.6	0.9
18.00	3070	42	318.6	5.0	6.5	0.9
19.00	3074	42	321.0	5.0	6.5	0.9
20.00	3074	47	322.9	5.0	6.4	1.0
21.00	3065	40	321.0	5.0	6.6	1.2
22.00	3072	42	421.7	5.0	6.5	1.4
23.00	3077	50	322.3	5.0	6.4	1.3
24.00	3086	30	321.7	5.0	6.3	1.2

Figure 13 shows the flow chart for obtaining the dynamic response of a rotor shaft. This is actualized by evaluation of the program code written in C++programming language from figure 12a. The program helped in the calculation of dynamic response by inputting the values obtained from GT 17 of Afam Power Station into the equation 40. The result shows that the rotor vibration response takes place at the same frequency as that of the excitation.

4. ANALYSIS

Gas Turbine, GT unit 17 is a healthy engine. The horizontal readings of natural frequency (ω_n) taken from healthy GT 17 and engine data are presented in Table 1, while the other parametric characteristics of the Afam GT 17 system used are shown in Appendix A. However, these readings were taken alongside the corresponding speed at different times for active and reactive loads.

4.1. Discussion

Table 1 shows the result of vibration displacement amplitude of bearings 1, 2 and 3 as a function of time. From the graph presented in figure 14, it was observed that bearing 1 and 2 started vibrating as soon as the engine was powered, while bearings 3 delayed for some operational interval before vibrating. The graph shown in figure 14 also shows that the machine tends to vibrate in higher displacement amplitude and sometimes slows down as the engine continues in operation (i.e. in fluctuating manner).

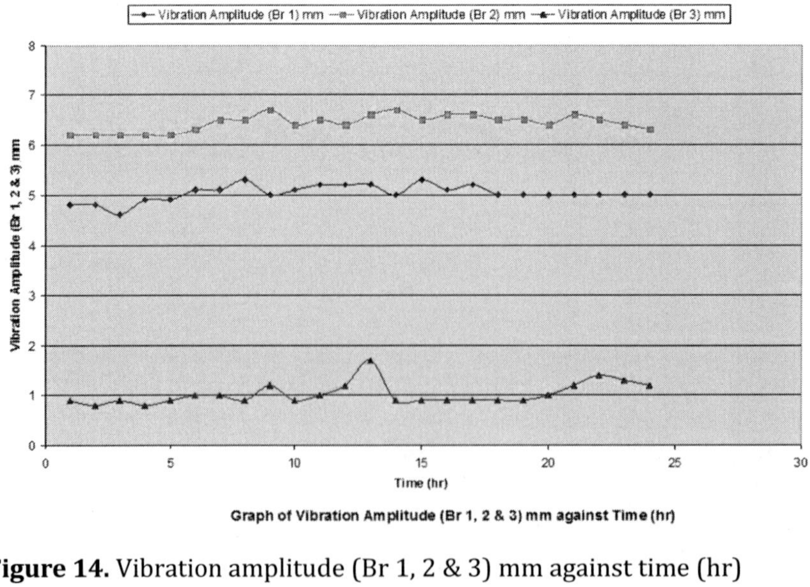

Figure 14. Vibration amplitude (Br 1, 2 & 3) mm against time (hr)

Depicted in figure 15 is the graph of the response of the system against time. Here, it is shown clearly that the forcing function x (t) is made up of series of impulses of varying magnitude. The impulse response was found to correspond with the sinusoidal shape as expected.

Figure 16 conversely shows the graph of natural frequency against time. From the graph, the result shows that the natural frequency tends to vary with time. It is observed also that the vibration of the engine occurs as a function of natural frequency at a given time.

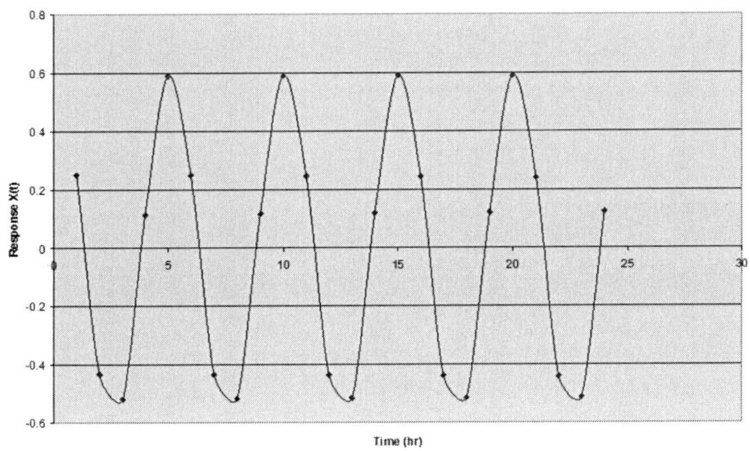

Figure 15. Response, X (t) against time (hr)

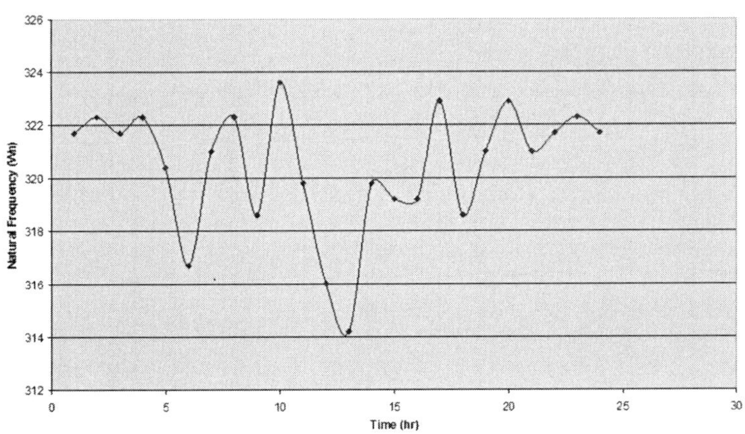

Figure 16. Natural frequency (ω_n) against time (hr)

Figure 17 further shows the graph of speed (rpm) with time. Hence, the sudden increase in speed as presented in the figure is capable of resulting to an increase in vibration at the given interval.

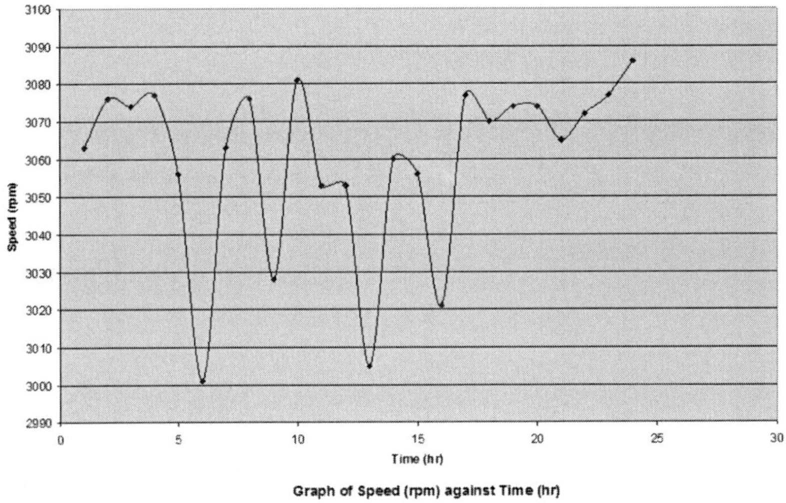

Figure 17. Speed (rpm) against time (hrs)

5. CONCLUSION

A work has been carried out on the modeling of dynamic response of marine GT rotor shaft systems. It is shown that when a system is subjected to force harmonic excitation, its vibration response takes place at the same frequency as that of the excitation.

To determine the response of the shaft under vibration, readings were collected from bearings 1, 2, and 3 of GT 17 in Afam thermal Station as shown in table 1, while the engine characteristics are shown in Appendix A. Equation 40 was developed to determine the response of the system under vibration. This mathematical equation is used to run a computer programme with a code in C++programming language.

6. RECOMMENDATIONS

The recommendations are as follows:

1. More attention should be paid to shaft vibrations as is the case with vibration on bearing.
2. Some factors which affect the performance of gas turbines on industrial duty should be considered while carrying out vibration based simulation of GT rotor shafts.
3. Errors and extraneous environmental factors should be put into consideration when modeling the response of rotor shaft under vibration.
4. GT rotor shaft systems should be provided with more supports to prevent adverse effects of eccentricity which leads to bow and whirl.

7. APPENDIX A

7.1. Characteristics of AFAM GT 17 system relevant to this work

Name of Equipment: Brown Boveri Sulzer Turbo Maschinen

Manufacturer:	Asea Brown Boveri
Capacity:	75 MW
Year of Manufacture:	1981
Year of installation	1982
Year of commissioning	Nov 1982
Type:	13D
No of Turbine Rows:	5
No. of Compressor Rows:	17

Particulars of Rotor

Length of Rotor shaft:	8000mm
Moment of Inertia, I:	$586.2m^4$
Modulus of Elasticity, E:	$207GN/m^2$
Mean Diameter, Θ:	1400mm
Density, ρ:	$7850kg/m^3$
Modulus of Rigidity, G:	$80GN/m^2$
Mass of Turbine shaft:	23500kg
Mass of Compressor shaft:	24000kg
Natural frequency, ω_n:	350rad/s
Damping Ratio, ζ ζ :	0.9
Spring stiffness, K:	$1.5x10^6N/mm$
Maximum vibration limit:	7.0mm/s

REFERENCES

1. Api 617 (2002) Axial and centrifugal compressor and expander-compressor for petroleum, chemical and gas industry services, 7th edition, Washington D.C, API publishing.

2. Bader K.A (2010) Rotor dynamic analysis requirements in API Standards with case studies.Proceedings of *ASME Turbo Expo*. 2010. Power for land and sea and air, June 14-181010 Glasgow Paper No. GT 2010-23127.

3. Choi, S.Y. and Mau (2009) Dynamitic analysis of geared rotor-bearing systems by the transfer matrix method.*ASME Design Engineering Technical Conferences*.(Part B), 84: 2967-2976.

4. Eshleman, R. and Eubanks, A (2007) On the critical speeds of a continuous rotor, *Journal of Engineering for Industry*, 91: 1180-1188.

5. Gupta, K. D.,Gupta K.and Athre, K (2003) Unbalance response of a dual rotor system: (theory and experiment) *Trans. Journal of Vibration and Acoustics*, 115: 427-435.

6. Harilirua and Srinivasan P.S. (2010). Vibration analysis of flexible coupling by considering unbalance, *Middle East Journal of Scientific Research* 5 (5): pp 336-345, 2010.

7. Hibner (2007) Dynamic response of viscous-damped multi-shaft jet engines. *Journal of Aircraft*, 12(4): 305-312.

8. Iwatsubo, S.Arii, and Kawai, R, 2009. Coupled lateral-torsional vibration of rotor system trained by gears, *Bulletin of ASME*, 27(224): 271-277.

9. Kahraman, A., Ozguven, H. N.,Houser D. R.and Zakrajsek, J. J, 2009.Dynamic analysis of geared rotors by finite elements, Transactions.*Journal of Mechanical Design*, 114: 507-514.

10. Kris, M., Marco, P and Koenraad De B, (2010) Rotor dynamic modeling as a powerful support tool for vibration analysi on large turbomachinery.*The* 8th*IFFoMM International Conference on rotor dynamics,* September 12-15, 2010/kist, seoulkora, pp 700-706.

11. Lee, W. and Ha, D.H. (2003) Coupled lateral and torsional vibration characteristics of a speed increasing geared rotor-bearing system, *Journal of Sound and Vibration*, 263(4): 725-742.

12. Lida., Tamura K and Kikuch, H (2009) Coupled torsional-flexural vibration of a shaft in a geared system of rotors (1st report), *Bulletin of the ASME*, 23(186): 2111-2117.

13. Mitchell, L.D. and Melleu, D.M. (2005) Torsional- lateral coupling in a geared high-speed rotor system, *ASME Design Engineering Technical Conferences*. 3 (Part B), 84(2): 977-989.

14. R. B., Bhat,T. S and Sankar (2009) Coupled torsional flexural vibration of a geared shaft system using finite element method, *Shock and Vibration Bulletin* (Part 3), 55: 13-25.

15. Ogbonnaya, E.A. (2004) *Modeling vibration base faults in Rotor Shaft of Gas Turbine*. Ph.D work Department of Marine Engineering, Rivers State University of Science and Technology, Port Harcourt, Rivers State, Nigeria, pp 64- 79.

16. Ogbonnaya, E.A., Johnson, K.T, Ugwu, H.U and Orji, C.U (2010) Component model-based condition monitoring of a Gas Turbine, *ARPN Journal of Engineering and Applied Sciences*, ISSN 1819-6680, Vol., No. 3, p 40

17. Ogbonnaya, E.A., Ugwu, H. U. and Diema, E.J (2013) A model-based mixed data approach for optimizing the performance of an offshore gas turbine compressor, Journal of Vibration Analysis, Measurement and Control, *Columbia International Publishing,* doi:10.7726/jvwp.2013.1001, pp30-43

18. Rao, J.S., Chang, J. R and Shiau, T. N2007. Coupled bending-torsion vibration of geared rotors, *ASME* Design Engineering Technical Conferences. 3 (Part B), 84-2: 977-989.

19. Rao, J.S, (2006) *Rotor Dynamics*, 3rd edition, New Age International Publishers, India.

20. Shiyu, Z. and Jianjun S. (2001) Active balancing and vibration control of rotating machine: *A survey journal of shock and vibration digest.* vol. 33, No. 4 July, 2001. Pp 361-371 (C) September, 2001 Saye Publication.

21. Surial, A and Kaushal A. (2008) Dynamic analysis of flexible turbo-rotor system using super elements.Rolls Roycc, Materials Polytechnics University Toronto Canada, pp 1-8.

22. Zhu, X, and Andres, S.L., (2007).Rotor dynamic performance of flexure pivot hydrostatic Gas bearing for oil free turbo machinery *ASME . Eng. Gas Turbines Power,* 129 pp 1020-1027.

9

CHAPTER

GAS TURBINE TRANSIENT PERFORMANCE TRACKING USING DATA FUSION BASED ON AN ADAPTIVE PARTICLE FILTER

Feng Lu [1,2,*], Yafan Wang [1], Jinquan Huang [1,2,*] and Yihuan Huang [1]

[1] *Jiangsu Province Key Laboratory of Aerospace Power Systems, Nanjing University of Aeronautics and Astronautics, Nanjing 210016, China*

[2] *Collaborative Innovation Center of Advanced Aero-Engine, Beijing 100191, China*

ABSTRACT

This paper considers the problem of gas turbine transient performance tracking in a cluttered environment. To increase the accuracy and robustness of state estimation, a data-fusion nonlinear estimation method based on an adaptive particle filter (PF) is proposed. This method needs local estimates transmitted to a central filtering unit for data fusion, and then global data feedback to the local PF for consensus propagation. The computational burden is shared by the local PF and central filtering unit in the data-fusion architecture. Furthermore, the PF algorithm used for the data fusion is embedded with the prior knowledge of engine health condition and adaptive to the measurement noise, and hence is called the adaptive PF. The heuristic information of state variables represented by inequality constraints tunes the local estimates by a probability density truncation method. The covariance of measurement noise is calculated by wavelet transform and utilized to update the particle importance function of the real time PF. The performance improvements of the proposed method are indicated through extensive experiments for gradual and abrupt shift performance tracking under conditions of gas turbine transient operation.

Keywords: gas turbine; performance tracking; data fusion; particle filter (PF); probability density truncation; wavelet transform

1. INTRODUCTION

Gas turbine engines provide the power for airplanes, and their reliability is vital to flight safety and performance. Nevertheless, the engine working conditions are terrible and they usually must endure high speeds, extreme temperatures and strong vibrations. Erosion and fouling of major components are unavoidable and result in a gradual deterioration of engine performance during its lifetime. Besides, foreign/domestic object damage will cause engine performance to sharply shift, and it is called an abrupt fault [1]. Engine data are periodically collected by airlines for engine health evaluation. Maintenance schedules adapted to reliable health tracking results leads to safe operation and reduced costs [2]. Hence, how to get reliable information about engine health conditions in time has drawn a lot of attention.

Engine health parameters, which are correction factors for the efficiency of the major components, are introduced to quantify the engine performance changes [3]. The engine performance tracking problem can be regarded as calculating the health parameters. Various methods such as Kalman filters (KFs), neural networks, fuzzy logic, genetic algorithms and expert systems have been proposed to obtain health parameters for engine health monitoring [3,4,5,6,7]. KF-based methods seem to be the most common ones for gas turbine health estimation, but these techniques are mainly focused on engine steady state, e.g., under cruise or average conditions [8].

The engine transient condition represents how a gas turbine operates from one steady state to another steady state when the input changes. The rotating components of the engine (fan, compressor, high-pressure turbine (HPT) and low-pressure turbines (LPT)) run close to their surge boundaries during transient behavior, and some performance parameters exceed their extreme value in a short time. Generally speaking, a low-bypass engine performance anomaly more easily occurs during a transient process, e.g., acceleration and deceleration. That is to say, gas turbine transient performance tracking is more urgent compared to steady state monitoring, but up to this point in time no studies of dynamic behavior monitoring for gas turbines have been presented. This paper proposes an adaptive particle filter (PF) based data fusion approach to engine transient health monitoring.

In this paper we emphasize the problem of monitoring the health performance of key gas turbine rotating components, including fans, compressors (high-pressure compressors (HPCs)), high-pressure turbines (HPTs) and low-pressure turbines (LPTs). Failures of engine actuators, sensors, and other components are not considered. The engine performance tracking of gradual drifts and abrupt shifts during transient operation is mainly addressed, but the technique can also be applied to engine steady behavior. A data fusion architecture to monitor engine health conditions is designed, and prior knowledge of engine health state and measurement noise is utilized for a Monte Carlo simulation to tune the state estimates. The present contribution is a derivation of a data fusion method based on adaptive PF for gas turbine transient performance tracking at the theoretical, implementation and performance level.

This paper is organized as follows: Section 2 presents a review of the basic PF and the problem formulation; the adaptive PF algorithm and data fusion architecture implementation for gas turbine health monitoring is given in Section 3; the performance comparisons of the proposed method for gradual and sharp shifts in different operation conditions are discussed in Section 4; and Section 5 concludes this paper.

2. PROBLEM FORMULATION AND PARTICLE FILTER

2.1. Problem Formulation

A gas turbine engine is a low-bypass turbofan engine, see Figure 1. A single inlet supplies airflow to the fan. Air leaving the fan is separated into two streams: one stream passes through the engine core, and the other stream passes through the annular bypass duct and then leaves. The fan is driven by the low pressure turbine. The air passing through the engine core moves through the HPC, which is driven by the HPT. Fuel is injected in the combustor and burned to produce hot gas for driving the turbines. The gas leaves the LPT and is mixed with the air from the bypass duct through the convergent nozzle, which has a variable cross section area.

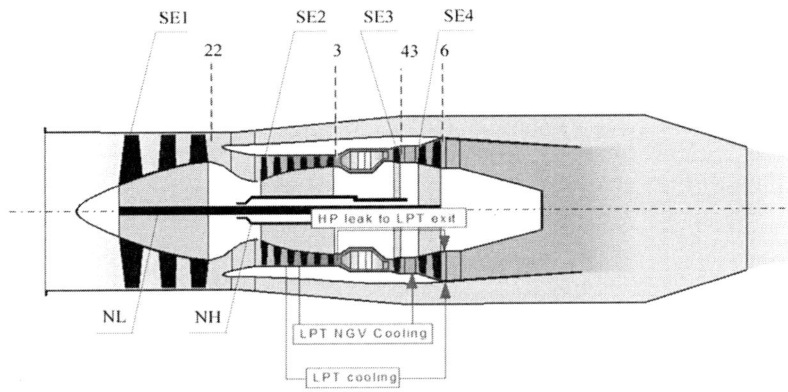

Figure 1. Schematic representation of a gas turbine engine.

Considering air flow mass, power and momentum conservation laws [9,10], a nonlinear aero-thermodynamic model of a gas turbine engine is given by:

$$x_{k+1} = f(x_k, u_k) + w_k$$

$$y_k = h(x_k, u_k) + v_k \tag{1}$$

where k is the time index, y is the 8-element measured output, x is the 6-element augmented state, and u is the 2-element control input. The noise terms w k and v k represent the process inaccuracies and measurement inaccuracies in the model. The sensor measurements are low-pressure spool speed NL, high-pressure spool speed NH, fan outlet pressure P22, HPC outlet pressure P3, fan outlet temperature T22, HPC outlet temperature T3, HPT outlet temperature T43 and LPT outlet temperature T6. The augmented state x includes the 2-element original state xo (NL and NH) and health parameter vector p (fan efficiency SE1, HPC efficiency SE2, HPT efficiency SE3 and LPT efficiency SE4, where section efficiency is defined SE). The elements of the control vector in the model are fuel flow Wf and nozzle area A8, which defines the engine operating point. There are two information entropy definitions used to select the system parameters. Auto Information Entropy is utilized to select the measured parameters, while the Cross Information Entropy to analyze the correlations between control, measured and health parameters.

In the framework of gas turbine performance monitoring, the quantities of interest are the differences between the estimated engine health parameters status and their reference ones. The actual engine performance is represented by the estimated values of health parameter, the prior value of which is adapted to the current measurements in a recursive approach. The block diagram of the model-based approach to monitoring the engine health condition is shown in Figure 2, and it is a closed-loop state estimator correcting structure. The KF, especially the linear KF (LKF), is an optimal state estimator for linear systems with noisy and inaccurate measurements, and is widely used for gas turbine engine health monitoring [11].

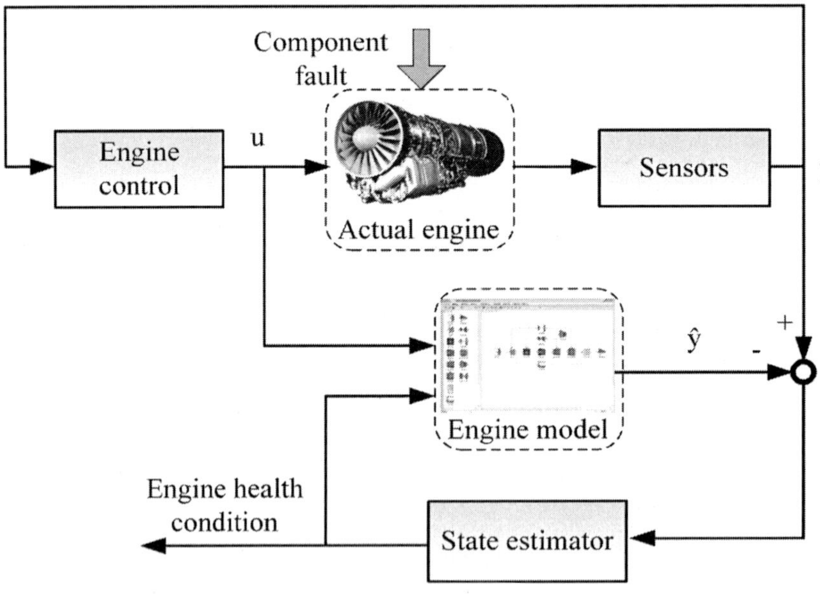

Figure 2. Gas turbine health monitoring based on a state estimator.

The state variable model (SVM) is a piecewise linear representation of the engine and needs to be established before the LKF implementation. It should be pointed out that the SVM obtained at a steady operating point is only a representation of the engine near this operating point. The state estimation accuracy by the LKF varies with the distance from the actual operating point to the design point, and also with the magnitude of the state deviation [12]. That means that the strong nonlinearity of gas turbine makes the LKF effective only in a small working range around the nominal state, and it can't do well in cases of transient behavior, e.g., during acceleration and deceleration.

Note that state variables change together in a situation of normal gradual engine deterioration, and the SVM is derived by the partial derivative calculation in the linearization process. Thus it is impossible for the SVM to correctly describe simultaneous shifts of all states, thus the tracking accuracy of gradual shifts by the LKF is not satisfactory. In addition, the assumption that the derivative of health parameter h is approximated by zero is utilized in the LKF equations, which is because engine deterioration generally occurs very slowly relative to the dynamics of the original state variables x0. Therefore any sharp shift of health parameters is not well estimated by the LKF due to the contradiction with the hypothesis $\dot{h}=0$. Hence, the generic LKF-based engine monitoring approach to the abrupt and multi-state gradual shift has a theoretical drawback.

To aim at the above issues, various nonlinear filtering methods are proposed and applied to state estimation, especially the extended Kalman filter (EKF), unscented Kalman filter (UKF) and PF [13,14]. Nevertheless, the EKF method is often used for weak nonlinear Gaussian systems due to a Taylor Series expansion truncated to the first order [8], and the performance closely depends on how often Jacobians are updated. Compared to the EKF and UKF, the PF is based on sequential Monte Carlo sampling theory, and it does not necessitate simplification of nonlinearity or any hypothesis of specific distributions [15,16]. Therefore, the PF-based approach is the one of the best ways that we address strongly nonlinear engine health monitoring issues in the following section.

2.2. The Particle Filter

The nonlinear model of a gas turbine is given by Equtaion (1), and let $x_{0:k} \triangleq \{x_0,\cdots,x_k\}$ and $y_{1:k} \triangleq \{y_1,\cdots,y_k\}$ denote the series of the state and measurement. Assume the probability density function (PDF) of the prior condition is $p(x_0)$, and the posterior PDF $p(x_{0:k}|y_{1:k})$ is characterized by a set of weighted random samples $\{x^i_{0:k},w^i_k\}^N_{i=1}$, wherein Ns is the particle number. The particle set $\{x^i_{0:k};i=0,\cdots,N\}$ is associated to the weights $\{w^i_k;i=0,\cdots,N\}$, and the PDF at time k can be approximated by:

$$p(x_{0:k}|y_{1:k}) \approx \sum_{i=1}^{N} w_k^i \delta(x_{0:k} - x_{0:k}^i)$$

$$\sum_{i=1}^{N} w_k^i = 1 \tag{2}$$

The case that the particles for Monte Carlo sampling are directly generated from the posterior PDF p(x 0:k |y 1:k) is expected, but it is generally unavailable. Thus the importance sampling distribution function q(x 0:k |y 1:k) is defined before sampling:

$$q(x_k | x_{0:k-1}, z_{1:k}) = q(x_k | x_{k-1}, z_k)$$

$$q(x_k^i | x_{k-1}^i, z_k) = p(x_k^i | x_{k-1}^i) \tag{3}$$

Then the ith particle weight w i k can be approximated by:

$$w_k^i \propto \frac{p(x_{0:k} | y_{1:k})}{q(x_{0:k} | y_{1:k})} \propto \frac{p(y_k | x_k^i) p(x_k^i | x_{k-1}^i) p(x_{0:k-1}^i | y_{1:k-1})}{q(x_k^i | x_{0:k-1}^i, y_{1:k}) q(x_{0:k-1}^i | y_{1:k-1})} = w_{k-1}^i \frac{p(y_k | x_k^i) p(x_k^i | x_{k-1}^i)}{q(x_k^i | x_{0:k-1}^i, y_{1:k})} \tag{4}$$

With this choice and normalization, the importance weights can be computed:

$$w_k^i \propto w_{k-1}^i p(y_k | x_k^i)$$

$$w_k^i = w_k^i / \sum_{i=1}^{N} w_k^i \tag{5}$$

One problem of the basic PF algorithm is that more particles have negligible weights after a few recursive steps. This implies that particle degeneracy occurs and a large computational effort for updating particle is meaningless. A typical method for solving this issue is importance re-sampling, and each particle is assigned by equal weight w i k =1/N whenever the effective number Neff of particles becomes less than a threshold value N th .

$$N_{\text{eff}} = \frac{1}{\sum_{i=1}^{N} \left(w_k^i\right)^2} < N_{\text{th}} \tag{6}$$

Once N th is close to value Neff,k, all particles have almost the same significance.

The architecture of a conventional PF mentioned above is that the measured data from different sensors are sent to one central PF to process together, and this is so-called the central architecture [15,17]. The advantage of this architecture is minimal information loss, but it also raises the problems that all measurements are identically treated at a time and the central filter bears a heavy computational burden, especially in the framework of the PF [18,19,20]. With the development of information fusion, the fusion PF structure employing a bank of local PFs and one master filter is presented and the computational effort is then shared by several filters [21,22].

In the case of the model-based approach to gas turbine health monitoring, the heuristic health knowledge is usually ignored and not considered in estimate algorithms. The typical anomaly modes and gradual deterioration rules of the engine have been summarized, and the magnitude ranges of health parameters are then determined. This prior knowledge about health parameters is represented by the constraints, and it can be used for state estimation in the PF algorithm. Furthermore, measurement noise levels vary with the order of the sensed value during the engine's dynamic operation. Generally speaking, the magnitude of sensed noise increases as the engine works at a larger operating point during the dynamic operation. Hence, it is very important for the PF to tune the estimate with the noise level in the situation of the transient behavior of the engine.

3. ADAPTIVE FUSION PARTICLE FILTER

3.1. The Adaptive Particle Filter

3.1.1. Particle Filter with Inequality Constraints

The prior state information can be described by equality or inequality constraints in the state estimation algorithm. These constraint approaches are mainly concentrated and applied to linear systems, especially combined with a KF, which has been proved to increase performance estimation accuracy. KFs with equality constraint approaches include the moving horizon estimation and smoothly constrained KF [23]. The estimate projection and truncation methods are incorporated to inequality constraints on state estimates, and the latter one has better performance for deterioration tracking due to its handling of two-sided constraints. The nonlinear state estimation with inequality constraints is more useful to gas turbine engine health monitoring, but there are not enough theoretical studies. For a nonlinear dynamic system, the heuristic knowledge represented by inequality constraints is introduced to the PF algorithm. This idea based on a previously published method [24] has been extended to the nonlinear PF approach for abrupt shift tracking.

The prior information we used in this paper is derived from the gradual deterioration rule and abrupt fault feature. Engine performance degrades with use, and the health parameters never improve and change in one way. For example, we know that the health parameter representing engine efficiency declines over time, and the parameter usually varies within the −10% magnitude due to abrupt faults. With this information, we can determine the bound of the state in advance and it is represented by the inequality constraint of the state estimator. The probability density function (PDF) of the PF estimate at the prior inequality constraint is truncated and the constrained filter estimate as the mean of the truncated PDF is calculated. Now consider the nonlinear system of Equation (1) where the s scalar constraints are added:

$$a_{k,m} \leqslant \phi_{k,m}^{\mathrm{T}} x_k \leqslant b_{k,m}, \qquad m = 1, \cdots, s \tag{7}$$

where $a_{k,m} \le b_{k,m}$, $a_{k,m}$ and $b_{k,m}$ are the two sided constraint of the mth health parameter at time k. The health parameter estimate $\hat{x}_k$ and covariance P_k is derived by the unconstraint PF algorithm:

$$\hat{x}_k = \sum_{i=1}^{N} w_k^i x_k^i$$

$$P_k = \sum_{i=1}^{N} w_k^i (x_k^i - \hat{x}_k)(x_k^i - \hat{x}_k)^{\mathrm{T}} \tag{8}$$

The PF algorithm with inequality constraints is to truncate the unconstrained PDF, $N(x_k, P_k)$. Once the mean $\tilde{x}_k$ and covariance $\tilde{P}_k$ of the truncated PDF are computed, we can obtain the constrained health parameter estimate. The state estimate $\tilde{x}_{k,m}$ and covariance $\tilde{P}_{k,m}$ are defined after the first m scalar constraints enforced, and then the new transformation is performed:

$$z_{k,m} = S_m W_m^{-\frac{1}{2}} T_m^{\mathrm{T}} (x_k - \tilde{x}_{k,m})$$

$$T_m W_m T_m^{\mathrm{T}} = \tilde{P}_{k,m}$$

$$S_m W_m^{\frac{1}{2}} T_m^{\mathrm{T}} f_{k,m} = \left[(f_{k,m}^{T} \tilde{P}_{k,m} f_{k,m})^{\frac{1}{2}} \quad 0 \quad L \quad 0 \right]^{\mathrm{T}} \tag{9}$$

where Tm is orthogonal, Wm is diagonal (these two quantities, Tm and Wm, can be derived from the Jordan canonical decomposition of P ˜ k,m) and Sm is obtained using Gram–Schmidt orthogonalisation. The first m inequality constraints are normalized:

$$c_{k,m} \leqslant [1 \;\; 0 \;\; 0 \cdots 0] z_{k,m} \leqslant d_{k,m}$$

$$c_{k,m} = \frac{a_{k,m} - {\phi_{k,m}}^{\mathrm{T}} \tilde{x}_{k,m}}{({\phi_{k,m}}^{\mathrm{T}} \tilde{P}_{k,m} \phi_{k,m})^{1/2}}$$

$$d_{k,m} = \frac{b_{k,m} - {\phi_{k,m}}^{\mathrm{T}} \tilde{x}_{k,m}}{({\phi_{k,m}}^{\mathrm{T}} \tilde{P}_{k,m} \phi_{k,m})^{1/2}} \tag{10}$$

Since zk,m is an identity covariance with statistically independent element, only the first element is constrained in Equation (10), and the PDF truncation reduces to a one-dimensional PDF. The part outside of the constraints is removed due to the fact zk,m lays between ck,m and dk,m. The truncated PDF is normalized after computing the area of the inside portion of the PDF. The mean μ m and variance σ 2 m of the first element of z ˜ k,m with the constraint enforcement are expressed by:

$$\begin{aligned}
\alpha &= \frac{\sqrt{2}}{\sqrt{\pi}(\mathrm{erf}(d_{k,m}/\sqrt{2}) - \mathrm{erf}(c_{k,m}/\sqrt{2}))} \\
\mu_m &= \alpha \textstyle\int_{c_{k,m}}^{d_{k,m}} \xi \exp(-\xi^2/2) d\xi = \alpha [\exp(-c_{k,m}^2/2) - \exp(-d_{k,m}^2(k)/2)] \\
\sigma^2{}_m &= \alpha \textstyle\int_{c_{k,m}}^{d_{k,m}} (\xi - \mu_m)^2 \exp(-\xi^2/2) d\xi \\
&= \alpha [\exp(-c_{k,m}^2/2)(c_{k,m} - 2\mu_m) - \exp(-d_{k,m}^2/2)(d_{k,m} - 2\mu_m)] + \mu_m^2 + 1
\end{aligned} \tag{11}$$

where α is a magnification factor. The inverse transformation of the Equation (9) is taken, and the mean and variance of the constrained state estimate are therefore given as:

$$\begin{aligned}
\tilde{x}_{k,m+1} &= T_m W_m^{1/2} S_m^{\mathrm{T}} z_{k,m+1} + \tilde{x}_{k,m} \\
\tilde{P}_{k,m+1} &= T_m W_m^{1/2} S_m^{\mathrm{T}} \tilde{C}_{k,m+1} S_m W_m^{1/2} T_m^{\mathrm{T}}
\end{aligned} \tag{12}$$

We repeat the process of Equations (9)–(12) to enforce the next constraint to the state estimate and jump out of the iteration until m = s. Hence, we obtain the state estimate and covariance at time k as x k =x˜ k,s ,P k =P˜ k,s by the constrained PF algorithm.

3.1.2. Measure Noise Tuned Particle Filter

As we know, particle importance weight wi is closely dependent on the distance between the real sensed value and its estimate, and the smaller the distance, the larger the weight. In the basic PF algorithm, the particle importance weight wi is defined as follows:

$$w_i = \frac{1}{\sqrt{2\pi\sigma}}\exp\left(-\frac{(y-\hat{y}_i)^2}{2\sigma^2}\right) \tag{13}$$

where the quantity σ is consistent with the covariance of measurement noise due to the fact that the sensed noise usually is generally recognized as Gaussian white noise. We set the quantity σ by the heuristic experience of the noise and it remains constant in the conventional PF algorithm. However, the probability distribution of the sensed noise changes in engine transient behavior, and it raises the problem of measurement noise uncertainty with regards to state estimation. The deviation from the actual measurement noise will result in a decline in estimation accuracy. In the paper the on-line tuning quantity σ of the PF with the measurement noise variance with the help of the wavelet transform is designed.

The measured stream y ω (k) could be approximated by a low-order polynomial or a piecewise low-order polynomial in an observation interval [25]. This interval size is corresponding to sampling number. Suppose the polynomial of the sensed stream y ω (k) is described as follows:

$$y_\omega(k) = a_0 + a_1 k + \cdots + a_M k^M + \omega \tag{14}$$

Then the wavelet transform is implemented to the sensed stream m w (k) :

$$W_{y_\omega}(s,\tau) = y_\omega(k) * \varphi_{s,\tau}(k) = W_y(s,\tau) + W_\omega(s,\tau)$$
$$\varphi_{s,\tau}(k) = \frac{1}{\sqrt{s}}\varphi\left(\frac{k-\tau}{s}\right) \tag{15}$$

where ϕ(k) is a wavelet function, "*" is the convolution operation, W y (s,τ) is an approximate part of the wavelet coefficients and W ω (s,τ) represents a detailed part of the wavelet coefficients. We define a vanishing moment α of the function ϕ(k) , and the measurement noise part is extracted from the sensed stream part once the quantity α is greater than the highest-order of the polynomial, namely α > M. Then the standard deviation of the noisy term in an interval can be calculated by the following expression:

$$\sigma = \frac{1}{M}\mathrm{Med}(|W_{y_\omega}(s,t_h)|) \tag{16}$$

where the scale s equals to 0.5, and the series $W_{y_\omega}(s,t_h)$ is the K/2 wavelet coefficients of $\{y_\omega(k)|k=0,1,\cdots K\}$, wherein $0\le h\le K/2$. Normalizing parameter M is usually set at 0.6745, and the function Med() represents the middle value calculation of a series.

In order to tune the quantity σ of the PF algorithm in real-time during the engine dynamic behavior, the measurement noise of a series in an interval is computed by a wavelet transformation. The sensed stream y_ω is segmented by a window with fixed width L, and it slides forward along the sampling time. The variance of the sensed noise part in the sliding window is estimated, therefore, the framework of the adaptive PF algorithm is established as shown in Figure 3.

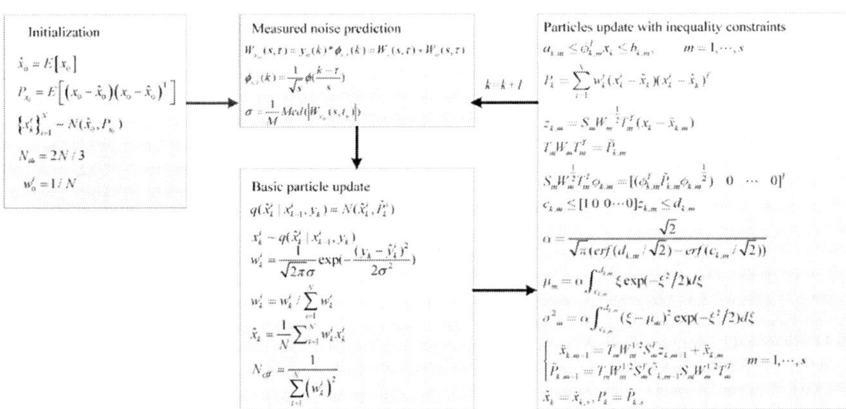

Figure 3. The framework of the adaptive particle filter (PF) algorithm.

3.2. Data Fusion Based on Adaptive Particle Filter

The fusing PF based on the integration of adaptive PF algorithm and information fusion theory is proposed for gas turbine transient performance monitoring. The sensor used to detect engine health condition is divided into several teams, and each team is applied to estimate the local estimates by the local PF. The engine component layout and thermodynamics of operation are taken into account for the partitions of local filters in the fusion architecture. The fan of the engine is a cold and low pressure (LP) component, the compressor is a cold and high pressure (HP) one, HPT is a hot and HP one, and LPT is a hot and LP one [26,27]. Two different ways to partitioning are performed, namely, the cold-hot component partition and the LP-HP component partition. Therefore, there are four combinations of the four rotating components above, which are the cold group, the hot group, the HP group and the LP group. For example, the cold partition includes two components (fan and compressor). Likewise, the hot partition consists of HPT and LPT, the HP partition of compressor and HPT, and the LP partition of fan and LPT.

For the cold group, the measured parameters are T22, P22, T3 and P3. In a similar manner the measurements T43 and T6 are for the hot group, the sensors T22, P22 and T6 for the LP group, and the sensors T3, P3, and T43 for the HP group. In addition, two spool speeds (NL and NH) are important quantities representing engine operation status, which are communal measurements and utilized in each part. Hence, the measurements of four local filter can be denoted by y1 = [NL, NH, T22, P22, T3, P3], y2 = [NL, NH, T43, T6], y3 = [NL, NH, T22, P22, T6], y4 = [NL, NH, T3, P3, T43]. Although the measurements of each engine component partition are different, the health parameters to be estimated are the same in every local filter, namely, SE1, SE2, SE3 and SE4.

The data fusion estimation based on the adaptive PF mainly includes three stages. First, several local filters perform in parallel to obtain individual sensor-based estimates. Second, all local estimates are combined in a master filter, where a global state estimate is yielded. Third, the global state and covariance is fed back to local filters with an information- sharing strategy for next cycle. The procedure of data fusion for the engine transient performance estimation is summarized as follows:

Step 1: Initialization

Given the initial values of global state x 0 , estimation error covariance P 0 , and process noise covariance Q 0 in the master filter. The four local filters are initialized with the information allocation strategy:

$$Q_{j,0} = \beta_j^{-1} Q_{m,0} \quad P_{j,0} = \beta_j^{-1} P_{m,0} \quad X_{j,0} = \beta_j^{-1} X_{m,0} \quad j = 1, \cdots, 4 \tag{17}$$

where the information distribution factor β j follows:

$$\sum_{j=1}^{4} \beta_j = 1 \tag{18}$$

Step 2: The adaptive PF performs in the local filter.

The particles {x i j,0:k−1 ,w i j,k−1 } N i=1 are generated based on the prior distribution N(X j,k−1 ,P j,k−1) , and are propagated through the nonlinear model Equation (1). The numerical characteristics of the measured noise are computed by wavelet transformation, and the quantity σ is applied to the importance weight calculation. The known health information is imposed to the PF to produce the local constrained estimates of the state x ^ j,k , error covariance P ^ j,k and noise covariance Q j,k .

Step 3: Information fusion implements in the master filter.

The local estimates x ^ j,k , P ^ j,k , and Q j,k are sent to the master filter to fulfill the information fusion for the global optimal estimate x m,k . The estimate error covariance P ^ j,k is an important parameter representing the performance of the local filter, and it is used to calculate the fusing weight of the local filter. The larger the covariance P ^ j,k is, the smaller the fusing weight in the global state is in the paper:

$$Q_{m,k} = (\sum_{j=1}^{4} Q_{j,k}{}^{-1})^{-1}$$

$$P_{m,k} = (\sum_{j=1}^{4} \hat{P}_{j,k}^{-1})^{-1}$$

$$x_{m,k} = P_{m,k} \times \sum_{j=1}^{4} \hat{P}_{j,k}^{-1} x_{j,k} \tag{19}$$

Step 4: Information distribution strategy

The state estimate calculated by the master filter is transmitted back to each local filter with an information assignment strategy:

$$Q_{j,k} = \beta_j^{-1} Q_{m,k} \quad P_{j,k} = \beta_j^{-1} P_{m,k} \quad X_{j,k} = X_{m,k} \quad j = 1 \cdots N \tag{20}$$

Due to the fact that information distribution factor β j has no effect on estimation accuracy, it is set by β j =1N .

Steps (2)–(4) present the fusing PF algorithm at iteration k. For iteration k + 1, the state and covariance delay a time index and Steps (2)–(4) are repeated. The fusion filter architecture is shown in Figure 4 for gas turbine transient performance estimation.

As shown in the figure, we can see that calculation loads are shared both by the local filters, and the master filter no longer undertakes the whole process like in the basic PF algorithm. Since the time update and measurement update are carried out independently in every local filter, the individual estimate by the local filter is not immediately affected by others. In the gas turbine engine health monitoring application, the data fusion filter architecture has a more efficient capability to deal with state estimation in cases of sensor fault due to fusing weight adaptive to estimation accuracy of the local filter.

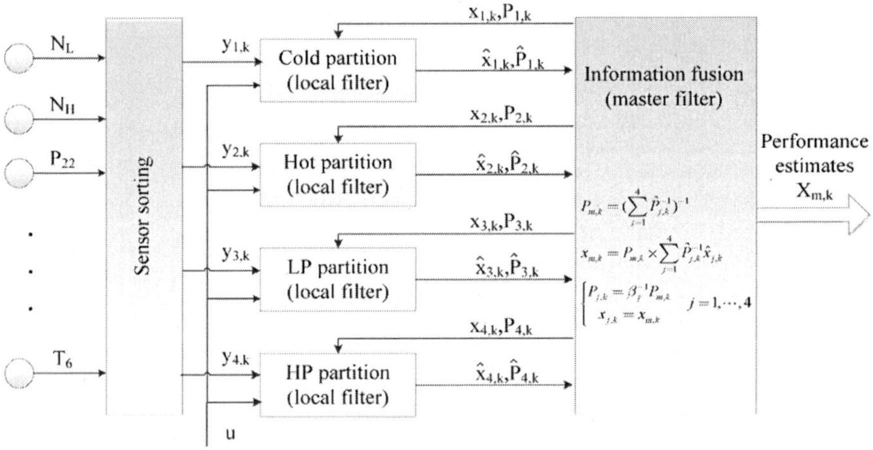

Figure 4. Data fusion filter architecture for gas turbine performance estimation.

4. SIMULATION AND ANALYSIS

The data fusion based on the adaptive PF approach is evaluated for the engine performance monitoring using the Matlab software. In the simulation environment, we use the component-level-model (CLM) engine to take the place of the actual engine, and the sampling rate equals to 50 Hz. The hardware of computer used for simulation is configured as follows: CPU i3-2100 @ 3.10 GHz and RAM 2GB. The standard deviations of measurement and health parameter are shown in Table 1 and Table 2. Gaussian noise v with magnitude specified in Table 1 is added to the simulated measured values, and the independent system noise and initial measured noise separately follow $\omega \sim N(0,Q)$ and $v \sim N(0,R)$, wherein $Q=0.16\times10^{-4} I_{6\times6}$.

Table 1. Gas turbine component-level-model (CLM) model measurements, nominal value and standard deviation. High-pressure compressor: HPC; high-pressure turbine: HPT; low-pressure turbine: LPT.

Measurement	Acronyms	Normalized Value	Standard Deviation
Low pressure spool speed	N_L	1	0.0015
High pressure spool speed	N_H	1	0.0015
Fan outlet temperature	T_{22}	1	0.002
Fan outlet pressure	P_{22}	1	0.0015
HPC outlet temperature	T_3	1	0.002
HPC outlet pressure	P_3	1	0.0015
HPT outlet temperature	T_{43}	1	0.002
LPT outlet temperature	T_6	1	0.002

Gas turbine health condition is represented by health parameters as mentioned in the previous section. Table 2 shows four abrupt faults representative of possible situations expected to be encountered in practice, and the health parameter deviation in each case refers to lab record of the Rolls-Royce Company (London, UK) [3,28]. For example, Case 1 is a fan abrupt fault with its efficiency SE1 deviating 1%. Assume that there are no fault on sensor measurements in the following experiments.

Table 2. Gas turbine engine abrupt fault modes and their deviation.

Scenarios	Acronyms	Fault Mode	Deviation	Standard Deviation
Case 1	*SE1*	Fan abrupt fault	−1% on *SE1*	0.0005
Case 2	*SE2*	HPC abrupt fault	−1% on *SE2*	0.0005
Case 3	*SE3*	HPT abrupt fault	−1% on *SE3*	0.0005
Case 4	*SE4*	LPT abrupt fault	−1% on *SE4*	0.0005

Engine gradual deterioration due to normal usage is simulated by linear drift of four health parameters, beginning from a healthy engine (the four parameters equal to 1) at cycle number n = 0 and with the degeneration at the end of the sequence at n = 6000: −2.18% on SE1, −6.71% on SE2, −3.22% on SE3 and −0.81% on SE4. Considering the magnitude of both the engine abrupt fault and deterioration, the bounds of the health parameter representing the PF inequality constraints are separately set by a = [1.005,1.005,1.005,1.005]T and b = [0.97,0.90,0.96,0.98]T. The performance of the engine anomaly detection is assessed by three indices, namely, root-mean-square error (RMSE), convergence time and root-mean-square deviation (RMSD). The RMSE and RMSD are separately defined by:

$$RMSE = [\frac{1}{S}\sum_{i=1}^{S}(\hat{x}_i - x_i)^2]^{\frac{1}{2}}$$

$$RMSD = [\frac{1}{S}\sum_{i=1}^{S}(\hat{x}_i - \bar{\hat{x}}_i)^2]^{\frac{1}{2}} \quad (21)$$

where S is the sampling step and x ^ ¯ i the mean of estimate value. The convergence time Tc is used to indicate the delay in fault recognition. We define this time index in this paper that is from the starting deviation to the estimate steady state within ±0.02% range and no longer out of this range in two consecutive steps.

4.1. Abrupt Fault Diagnosis in Steady Operation Conditions

The tests on gas turbine abrupt fault diagnosis are first performed at ground steady conditions (H = 0 m, Ma = 0, Wf = 2.48 kg/s). The abrupt faults depicted in Table 2 are simulated, and the noise is not changed in the engine steady behavior. Given the stochastic character of the measurement noise, each test-case has been run five times. Then the engine fault diagnostic performances of basic KF, basic PF, fusion particle filter (F-PF) and fusion adaptive particle filter (FA-PF) are given in Table 3. The particle number of the PF is 60, and that of both the F-PF and FA-PF is 30.

The RMSEs and RMSDs of the two fusion PF approaches shown in Table 3 are smaller than those of the conventional PF and KF in the cases of the four abrupt fault modes, and the FA-PF one is superior to the others. The convergence time of the three PF approaches are nearly the same, and vary clearly in different cases. The KF is a linear estimator and it takes less time to reach the steady state. Two speed measurements are repeatedly utilized in each local filter and there is the reason for the fact the importance of their weights is increasing compared to the remaining measurements in the fusion filter structure. The fusion PF approaches have more satisfactory estimation accuracy due to sufficient extraction and information-sharing of key measurements such as the speeds for health monitoring. Because of the heuristic health knowledge enforced through inequality constraints, the FA-PF has the best estimation accuracy of fault diagnosis in the engine steady operation.

Table 3. The engine fault diagnostic performances of four filtering approaches in steady behavior. Kalman filter: KF; fusion particle filter: F-PF; fusion adaptive particle filter: FA-PF.

Fault Modes	Root-Mean-Square Error (*RMSE*)				Root-Mean-Square Deviation (*RMSD*)				T_c (ms)			
	KF	PF	F-PF	FA-PF	KF	PF	F-PF	FA-PF	KF	PF	F-PF	FA-PF
Case 1	0.0141	0.0108	0.0073	0.0059	0.0090	0.0085	0.0052	0.0044	220	190	230	220
Case 2	0.0137	0.0111	0.0078	0.0057	0.0094	0.0089	0.0059	0.0047	260	440	420	440
Case 3	0.0113	0.0118	0.0087	0.0060	0.0093	0.0096	0.0061	0.0050	320	620	660	600
Case 4	0.0126	0.0115	0.0080	0.0067	0.0100	0.0091	0.0058	0.0051	460	680	760	640

4.2. Abrupt Fault Diagnosis in Dynamic Operation

In order to further evaluate the proposed method performance in engine transient performance tracking, more simulation is carried out in the case of mixed gradual deterioration and abrupt faults. Gradual deterioration refers to all health parameters degrading linearly from the healthy condition to the end of 3000 cycles, and a simulated abrupt fault of magnitude −1% is added to SE2 at 5 s. The engine operates in the dynamic behavior mode from Wf = 2.48 kg/s to 1.98 kg/s under three operation conditions: ground (H = 0 m, Ma = 0), high-altitude 1 (H = 8000 m, Ma = 0.5), and high-altitude 2 (H = 11000 m, Ma = 0.8). The particle number of the three PFs is set the same as previously, and the measurement noises remain unchanged. A comparison of the three approaches for the engine transient performance tracking in the cases of mixing gradual deterioration and abrupt fault at ground is depicted in Figure 5, where the dotted line and solid line are the real and estimated values of health parameters, respectively.

It can readily be found from the Figure 5 that the PF working in fusion architecture outperforms the basic PF. Table 4 further presents the estimation performance by the three PF methods in terms of number and data at different operating conditions. As can be seen from Table 4, the RMSE of the F-PF and FA-PF is smaller than that of PF, among which the FA-PF in three operation conditions are below 0.007. Hence, the estimation accuracy of fusion PF architecture outperforms that of the generic PF structure.

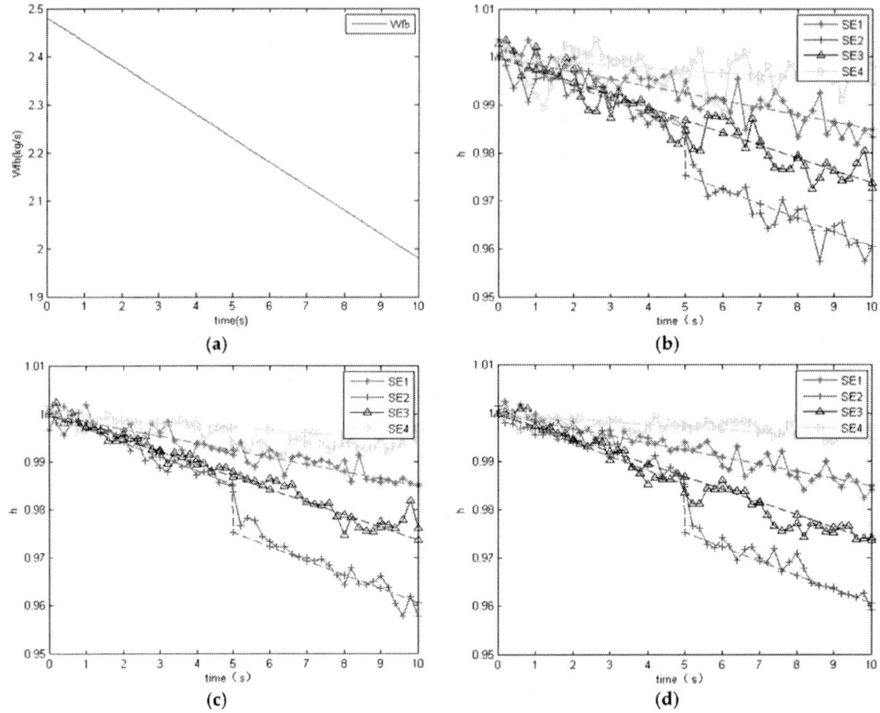

Figure 5. Engine transient performance tracking at ground condition. (a) Fuel supply rule Wf; (b) the performance estimates by the PF; (c) the performance estimates by the F-PF; and (d) the performance estimates by the FA-PF.

Table 4. The RMSE of dynamic estimation at three operation conditions.

Operation Condition	PF	F-PF	FA-PF
Ground	0.0120	0.0077	0.0069
High-altitude 1	0.0124	0.0076	0.0069
High-altitude 2	0.0123	0.0079	0.0070

4.3. Performance Estimation with Uncertain Noise in Dynamic Operation

The stochastic feature of the engine measured noise changes at different operation points. An experiment of gradual deterioration tracking with uncertain measurement noise is performed to assess the FA-PF algorithm. The engine experiences dynamic operation in the case of gradual deterioration as the same as the Section 4.2, but no abrupt fault is added. A series of the tests are implemented with ground conditions, including the change of only one measurement noise and all measurement noises simultaneously. Noise generated in the engine core, by sources such as the HPC, combustor, HPT and LPT, plays the most important roles to the overall noise under low-power conditions. While jet and fan noises have dominated over core noise at high engine power during takeoff [29].

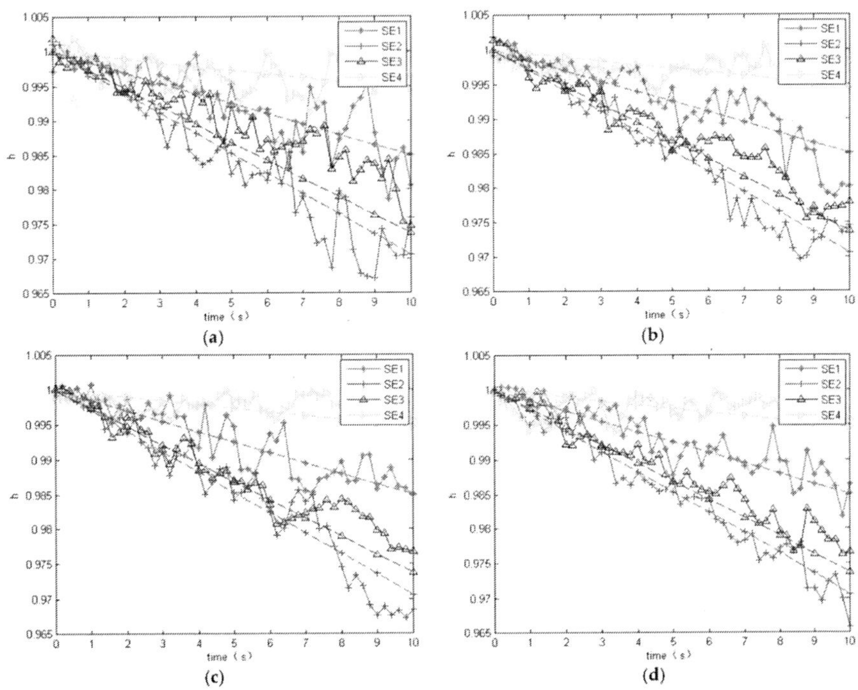

Figure 6. Transient performance monitoring with uncertain noise of one sensor P22 in the case of gradual deterioration. (a) σ = R0; (b) σ = R1; (c) σ = R2; and (d) adaptive σ.

The sensor noise varies with engine power condition in transient process. The uncertain noise of one sensor P22 in Figure 6 is simulated by the route of R0 = diag[0.0015, 0.0015, 0.002, 0.0015, 0.002, 0.0015, 0.002, 0.002] in 0–3 s, R1 = diag[0.0015, 0.0015, 0.002, 0.003, 0.002, 0.0015, 0.002, 0.002] in 3.02–6 s and R2 = diag[0.0015, 0.0015, 0.002, 0.0045, 0.002, 0.0015, 0.002, 0.002] in 6.02–10 s. The uncertain noise of each measurement undertakes the varied noise by R0 in 0–3 s, 2R0 in 3.02–6 s, and 3R0 in 6.02–10 s in Figure 7. The size of the sliding window by wavelet transform is 50 steps.

As shown in Figure 6, the engine transient performance estimates by the fusion PF approaches seem to deviate from their real values once the noise of sensor P22 varies, no matter the quantity of measurement noise σ is equal to R0, R1 or R2. The fusion adaptive PF has sound tracking performance due to the quantity σ adaptive to the real measurement noise real time in Figure 6d. We can obtain similar results in the case that all sensors for transient performance monitoring have uncertain measured noise in Figure 7.

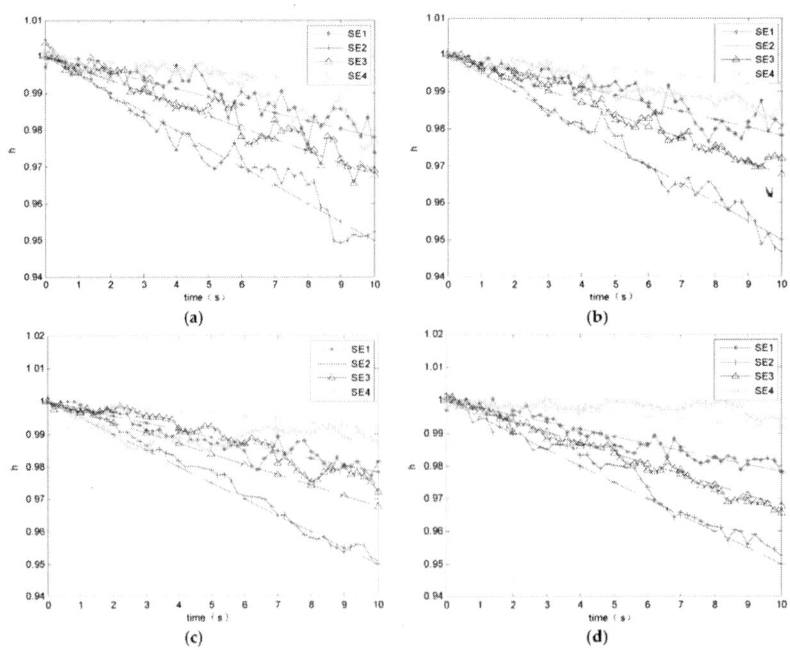

Figure 7. Transient performance monitoring with uncertain noise of every sensor in the case of gradual deterioration. (a) σ = R0; (b) σ = 2R0; (c) σ = 3R0; and (d) adaptive σ.

Table 5 summarizes the RMSE of the engine transient health estimation performance by the fusion PF approach with different the variable σ. The covariance of the measurement noise is estimated and real time tunes the quantity σ of the FA-PF algorithm as presented previously, and the RMSEs of the FA-PF are almost the same as those of the fusion PF where quantity σ is equal to the true value of measurement noise covariance in Table 5. Nevertheless, if the quantity σ is not tuned to the covariance of the real measurement noise, the fusion PF will produce a larger RMSE in terms of the engine transient performance estimation.

Table 5. The RMSE of dynamic estimation with uncertain measurement noise at ground.

σ	Uncertain Noise of Sensor P_{22}				
	R_0	R_1	R_2	True Value	Tuning Value
RMSE	0.0121	0.0109	0.0108	0.0085	0.0088
σ	Uncertain Noise of All Sensors				
	R_0	$2R_0$	$3R_0$	True Value	Tuning Value
RMSE	0.0165	0.0116	0.0169	0.0095	0.0096

4.4. Engine Health Monitoring Test

Finally a test of the engine health monitoring is carried out to evaluate the proposed method at ground level. The engine input variables, Wf and A8, are fed into the engine as shown in Figure 8. The engine NH representing the engine operation varies as follows: about 0.91 before 2.7 s, increasing from 0.91 to 1.0, then decreasing from 1.0 to 0.91, and about 0.91 to the end. During this process, the engine thrust increases from 0.819 to 1.0, and then back to 0.819. Four abrupt faults depicted in Table 2 are separately injected into the engine at 2 s.

Figure 9 depicts the variance estimates of measurement noise, which are applied to tune the quantity σ in the FA-PF method. The noise pollutes the true measurement, and it changes along the engine operation condition (the larger the power condition, the more noise enforcement). Table 6 summarizes the performance of engine health monitoring by four PF methods in the abrupt fault cases. The fusion PF with inequality constraints is defined by the FC-PF. The detection results in Table 6 show that the FC-PF and FA-PF have the less estimation errors and provide more stable estimates than the other PF methods because of the prior knowledge used. Furthermore, the performance of the FA-PF for engine health monitoring is the best due to the quantity of the importance function σ adaptive to the measurement noise real-time.

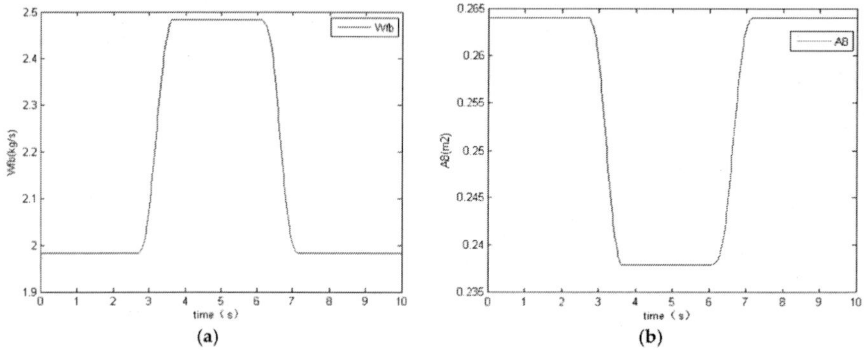

Figure 8. The change rules of the engine input variables. (a) Wf; and (b) A8.

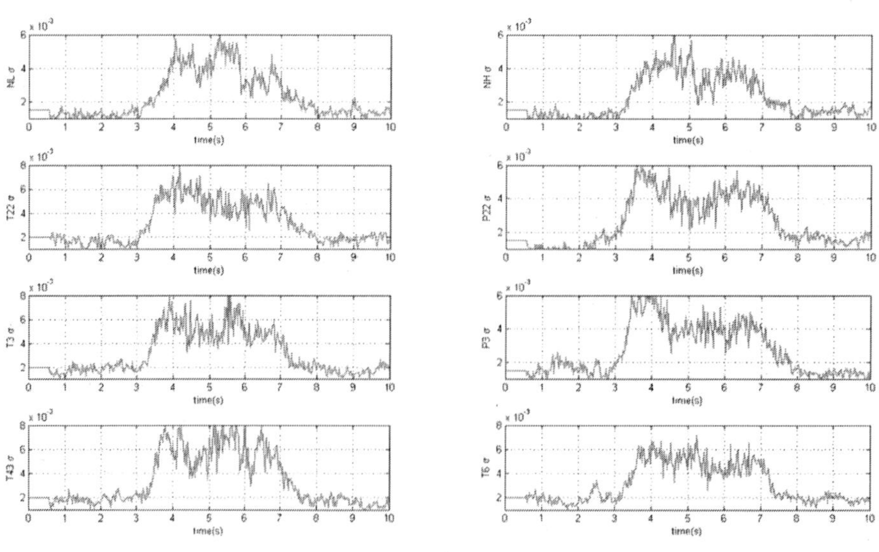

Figure 9. The variance estimates of measurement noise in the FA-PF.

Table 6. Engine health monitoring performance by four PF methods in the cases of abrupt fault.

Fault Modes	*RMSE*				*RMSD*			
	PF	F-PF	FC-PF	FA-PF	PF	F-PF	FC-PF	FA-PF
Case 1	0.0140	0.0101	0.0084	0.0061	0.0123	0.0083	0.0075	0.0050
Case 2	0.0138	0.0095	0.0078	0.0055	0.0119	0.0079	0.0069	0.0043
Case 3	0.0134	0.0100	0.0089	0.0058	0.0117	0.0069	0.0079	0.0050
Case 4	0.0144	0.0106	0.0094	0.0065	0.0132	0.0077	0.0079	0.0052

5. CONCLUSIONS

This paper describes the use of data fusion based on the adaptive PF for gas turbine dynamic performance monitoring. The state estimation in the fusion estimator architecture includes three steps: several local filters working in parallel to obtain individual sensor-based estimates, one master filter fusing these local estimates to yield a global state estimate, and the global estimates serving as the feedback to each local filter with information-sharing strategy. A systematic comparison of the fusion PF methods for transient performance estimation is presented. Gradual deterioration, abrupt faults and their mixtures are typically considered as the engine anomaly scenarios in the test. The fusion PF architecture has better estimation accuracy and less convergence time than the conventional PF architecture. The implementation of the fusion PF is quite straightforward and involves only basic matrix operations. The convergence time by the fusion PF is similar to that by the basic PF, yet the computational burden of the master filter is reduced, because it is shared by local filters in the fusion PF structure.

Moreover, an adaptive PF algorithm is proposed to sufficiently utilize prior information for the engine transient performance detection with measurement noise uncertainty. The heuristic health knowledge is usually neglected in model-based engine diagnoses due to the complex mathematics application to the conventional PF. In this paper, the engine prior health information represented by the inequality constraints is enforced in the fusion PF algorithm. In addition, the uncertainty of the measurement noise in the engine dynamic operation is considered in the fusion PF. The covariance of the sensed noise is estimated and then applied to tune the importance weights of the PF. The improvements brought by the data fusion based on the adaptive PF have been illustrated on the application of the engine transient performance monitoring with noise uncertainty. The experiments show that the proposed method leads to the more reliable assessments of the engine health conditions no matter whether the cases involve gradual engine deterioration or abrupt faults. With information fusion, a priori knowledge and measurement noise adaption in mind, the data fusion based on the adaptive PF approach seems therefore to be more promising. The present work by the authors has shown the advantages of data fusion based on an adaptive PF for gas turbine health monitoring. It is not considered that the phase errors between different measurements combined with the senor lag will affect the shift in engine performance, and it would be interesting to discuss the proposed approach of this paper with addition of measurement differences.

ACKNOWLEDGMENTS

We are grateful for the financial support of the National Nature Science Foundation of China (No. 61304133), the Fundamental Research Funds for the Central Universities (No. NS2015024). Moreover, the authors wish to thank the anonymous reviewers for their constructive comments and great help in the writing process, which improve the manuscript significantly.

AUTHOR CONTRIBUTIONS

Feng Lu and Jinquan Huang contributed in developing the ideas of this research, Yafan Wang and Yihuan Huang performed this research. All of the authors were involved in preparing this manuscript.

REFERENCES

1. Volponi, A. Gas turbine engine health management: Past, present, and future trends. J. Eng. Gas Turbines Power 2014, 136.
2. Rodger, J.A. Toward reducing failure risk in an integrated vehicle health maintenance system: A fuzzy multi-sensor data fusion Kalman filter approach for IVHMS. Expert Syst. Appl. 2012, 39, 9821–9836.
3. Borguet, S.; Léonard, O. Comparison of adaptive filters for gas turbine performance monitoring. J. Comput. Appl. Math. 2010, 234, 2202–2212.
4. Li, Y.G.; Korakianitis, T. Nonlinear weighted-least-squares estimation approach for gas-turbine diagnostic applications. J. Propuls. Power 2011, 27, 337–345.

5. Joly, R.B.; Ogaji, S.O.T.; Singh, R.; Probert, S.D. Gas-turbine diagnostics using artificial neural-networks for a high bypass ratio military turbofan engine. Appl. Energy 2004, 78, 397–418.

6. Vanini, Z.N.S.; Khorasani, K.; Meskin, N. Fault detection and isolation of a dual spool gas turbine engine using dynamic neural networks and multiple model approach. Inf. Sci. 2014, 259, 234–251.

7. Eustace, R.W. A real-world application of fuzzy logic and influence coefficients for gas turbine performance diagnostics. J. Eng. Gas Turbines Power 2008, 130.

8. Simon, D. A comparison of filtering approaches for aircraft engine health estimation. Aerosp. Sci. Technol. 2008, 12, 276–284.

9. Sun, J.G.; Vasilyev, V.; Ilyasov, B. Advanced Multivariable Control Systems of Aeroengines; Beihang Press: Beijing, China, 2005; pp. 60–83.

10. Lu, F.; Chen, Y.; Huang, J.Q.; Zhang, D.D. An integrated nonlinear model-based approach to gas turbine engine sensor fault diagnostics. J. Aerosp. Eng. 2014, 228, 2007–2021.

11. Volponi, A. Enhanced Self Tuning On-Board Real-Time Model (eSTORM) for Aircraft Engine Performance Health Tracking; Technical Report for National Aeronautics and Space Administration: Cleveland, OH, USA, 2008.

12. Armstrong, J.B.; Simon, D.L. Implementation of an Integrated On-Board Aircraft Engine Diagnostic Architecture; Technical Report for National Aeronautics and Space Administration: Cleveland, OH, USA, 2012.

13. Simon, D. Kalman filtering with state constraints: A survey of linear and nonlinear algorithms. IET Control Theory Appl. 2010, 4, 1303–1318.

14. Lu, F.; Huang, J.Q.; Lv, Y.Q. Gas path health monitoring for a turbofan engine based on a nonlinear filtering approach. Energies 2013, 6, 492–513.

15. Gordon, N.; Salmond, D.; Smith, A. Novel Approach to Nonlinear/Non-Gaussian Bayesian State Estimation. IEE Proc. F Radar Signal Process. 1993, 140, 107–113.

16. Climente-Alarcon, V.; Antonino-Daviu, J.A.; Haavisto, A.; Arkkio, A. Particle filter-based estimation of instantaneous frequency for the

diagnosis of electrical asymmetries in induction machines. IEEE Trans. Instrum. Meas. 2014, 63, 2454–2463.

17. Zhao, B.; Skjetne, R.; Blanke, M.; Dukan, F. Particle filter for fault diagnosis and robust navigation of underwater robot. IEEE Trans. Control Syst. Technol. 2014, 22, 2399–2407.

18. Tao, G.L.; Deng, Z.L. Self-tuning fusion Kalman filter for multisensor single-channel ARMA signals with coloured noises. IMA J. Math. Control Inf. 2015, 32, 55–74.

19. Zhu, H.Y.; Zhai, Q.Z.; Yu, M.W.; Han, C.Z. Estimation fusion algorithms in the presence of partially known cross-correlation of local estimation errors. Inf. Fusion 2014, 18, 187–196.

20. Seifzadeh, S.; Khaleghi, B.; Karray, F. Distributed soft-data-constrained multi-model particle filter. IEEE Trans. Cybern. 2015, 45, 384–394.

21. Zajac, M. Online fault detection of a mobile robot with a parallelized particle filter. Neurocomputing 2014, 126, 151–165.

22. Li, T.C.; Sun, S.D.; Sattar, T.P. Fight sample degeneracy and impoverishment in particle filters: A review of intelligent approaches. Expert Syst. Appl. 2014, 41, 3944–3954.

23. Simon, D. Constrained Kalman Filtering via Density Function Truncation for Turbofan Engine Health Estimation; Technical Report for National Aeronautics and Space Administration: Cleveland, OH, USA, 2006.

24. Boulkroune, B.; Darouach, M.; Zasadzinski, M. Moving horizon state estimation for linear discrete-time singular systems. IET Control Theory Appl. 2010, 4, 339–350.

25. Yadav, S.K.; Sinha, R.; Bora, P.K. Electrocardiogram signal denoising using non-local wavelet transform domain filtering. IET Signal Process. 2015, 9, 88–96.

26. Kyriazis, A.; Mathioudakis, K. Enhance of fault localization using probabilistic fusion with gas path analysis algorithms. J. Eng. Gas Turbines Power 2009, 131.

27. Kaltungo, A.Y.; Sinha, J.K.; Elbhbah, K. An improved data fusion technique for faults diagnosis in rotating machines. Measurement 2014, 58, 27–32.

28. Curnock, B. Obidicote Project—Work Package 4: Steady-State Test Cases; Rolls-Royce PLC: Manchester, UK, 2000.

29. Hultgren, L.S.; Miles, J.H. Noise-Source Separation Using Internal and Far-Field Sensors for a Full-Scale Turbofan Engine. In Proceedings of the 15th AIAA/CEAS Aeroacoustics Conference (30th AIAA Aeroacoustics Conference) Cosponsored by AIAA and CEAS, Miami, FL, USA, 11–13 May 2009.

10

CHAPTER

THE IMPORTANCE OF HOT CORROSION AND ITS EFFECTIVE PREVENTION FOR ENHANCED EFFICIENCY OF GAS TURBINES

I. Gurrappa[1], I.V.S. Yashwanth[2], I. Mounika[1], H. Murakami[3] and S. Kuroda[3]

[1] Defence Metallurgical Research Laboratory, Kanchanbagh PO, Hyderabad, India

[2] University of Louisiana, Lafayette, LA, USA

[3] National Institute for Materials Science, Sengen, Tsukuba City, Ibaraki, Japan

1. INTRODUCTION

Advanced gas turbine engines essentially need novel materials to exhibit forever-greater efficiency by increasing their operating temperatures. Higher operating temperatures lead to high temperature corrosion of the components and thereby reduce their life significantly. The blades in modern aero, marine and industrial gas turbines are manufactured primarily from Ni - based superalloys. Oxidation and hot corrosion are the two important factors (under high temperature corrosion), which determine the life of gas turbine engine components. The rate of degradation is slow under oxidation conditions, while it is significantly fast under hot corrosion conditions and catastrophic failures are imminent if proper materials in association with appropriate coatings are not used. Hot corrosion is major problem for marine and industrial gas turbines while aero gas turbines experience it while they move across the sea. Assessment of current status on hot corrosion problems in different gas turbine engines is imperative in order to improve their life by selecting advanced hot corrosion resistant materials and coatings. This process helps in highlighting the issues that need to be addressed not only for enhancing the efficiency of gas turbine engines but also to avoid failures during service.

A section of a typical gas turbine engine is shown in Fig.1 [1]. It is important to note that hot section components of the engine are manufactured from Ni-based superalloys. Advances in processing of Ni-based superalloys have allowed evolution of microstructures from equiaxed structures about three decades ago to directionally solidified (DS) multi-grain and single crystal (SC) components today. With added capability from compositional flexibility coupled with advances in processing over that time period, high-pressure turbine blade temperatures have increased marginally and metal surface temperature at the hottest locations approach 1150^0 C in state-of-the–art gas turbine engines. The most severe combinations of stress and temperature in the present engine designs correspond to bulk average metal temperatures approaching 1000^0 C using the first generation single crystal alloys.

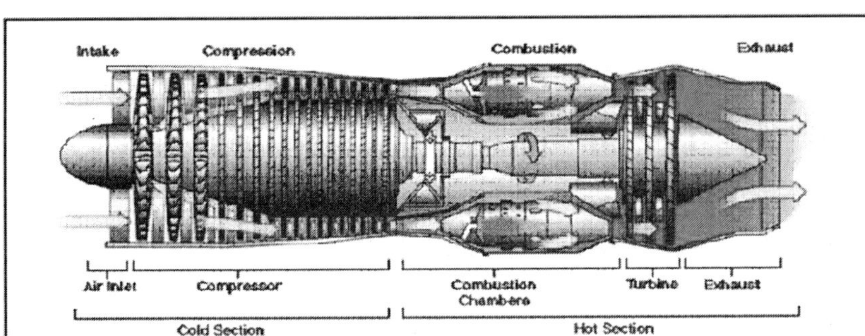

Figure 1. Gas turbine engine showing the usage of superalloys and titanium based materials [1]

One of the most critical components in the engine is the gas turbine blade. The high-pressure turbine blade operates under the most arduous conditions of temperature and stress than any component in the engine.

Not only does the blade experience high temperature and direct stress, it also experiences rapid temperature transients at various points during the engine cycle. The hot gases surrounding the blade are highly oxidizing and contain high levels of contaminants like sulphur and chlorine if low-grade fuels are used. An ideal superalloy / coating should be able to survive this harsh corrosive environment.

The efficiency of a gas turbine is proportional to firing or turbine inlet temperature. Increase in engine operating temperature has meant that the traditional corrosion resistant turbine blade alloys such as IN 738 and IN 939 are no longer strong enough to last the expected 25,000 hours minimum life [2]. It implies that higher strength alloys like Mar M247, CM247 LC, single crystal alloys, CMSX-4, CMSX-10 etc. are required for creep strength. Progress in aero and industrial gas turbine blade materials is illustrated in Fig.2 [2]. During the last one decade, dramatic competition in the power equipment industry has boosted technology to the levels achieved in the aviation turbines just years before. Latest industrial gas turbines use single crystal, rhenium containing Ni-based superalloys and directionally solidified blades and vanes. These alloys have relatively poor hot corrosion resistance and hence have to rely on advanced protective coatings to prevent severe and potentially life limiting damage.

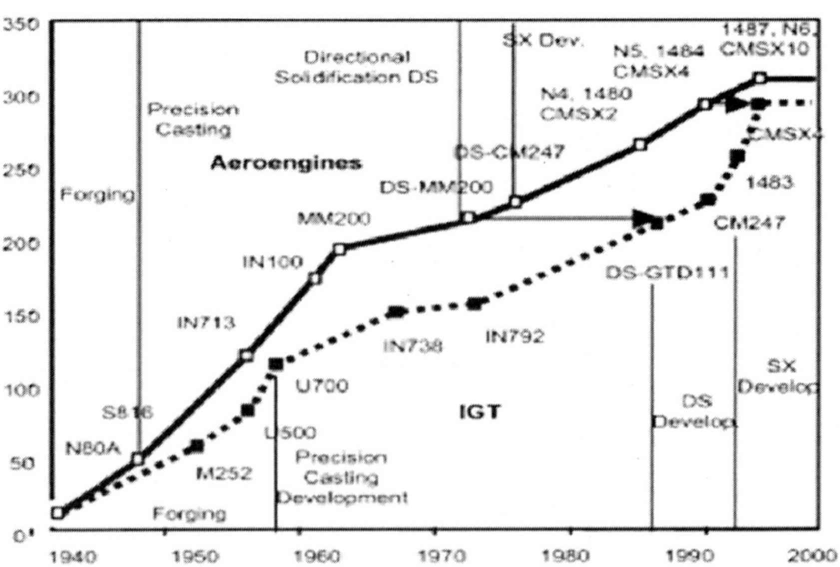

Figure 2. Progress in aero and industrial gas turbine blade superalloys [2]

The majority of Ni - based superalloy development efforts has been directed towards improving the alloy high temperature strength with relatively minor concern being shown to its hot corrosion resistance. Further, it is not always possible to achieve both high temperature strength and hot corrosion resistance simultaneously because some alloying elements help to improve hot corrosion resistance while some may help to improve high temperature strength. It is rare that an alloying element leads to enhancement both in high temperature strength and the hot corrosion resistance. This is further complicated for marine applications by the aggressivity of the environment, which includes sulphur and sodium from the fuel and various halides contained in seawater. These features are known to drastically reduce the superalloy component life and reliability by consuming the material at an unpredictably rapid rate, thereby reducing the load-carrying capacity and potentially leading to catastrophic failure of components [3-5]. Thus, the hot corrosion resistance of superalloys is as important as its high temperature strength in gas turbine engine applications. Recent studies have shown that the high temperature strength materials are most susceptible to hot corrosion and the surface engineering plays a key role in effectively combating the hot corrosion problem [6-7]. Therefore, the two materials (base and coating) must be considered as an integral system and the interface between them is also equally important as it can often be the limiting factor.

This chapter explains the critical issue of hot corrosion of superalloy components in gas turbine engines and methodologies to select appropriate materials and coatings for its prevention. It is followed by an assessment of current status of coatings, coating techniques and finally, some of the critical areas that need to be addressed for development of more efficient and smart coatings.

2. HOT CORROSION

Hot corrosion became a topic of important and popular interest in the late 60s as gas turbine engines of military aircraft suffered severe corrosion during the Vietnam conflict during operation over seawater. Metallographic inspection of failed components showed sulphides of nickel and chromium and hence the process was named ‘sulphidation’. However, the process has subsequently been renamed “hot corrosion” because corrosion by a thin electrolyte film bears some common features with 'atmospheric corrosion' by an aqueous film at room temperature.

Break down of protective oxide layers by chemical interaction with certain aggressive species contained in the combustion environment can produce accelerated attack on the underlying metal. This process is known as hot corrosion. Hot corrosion takes place mainly because of high concentrations of sulphur, vanadium and sodium in the fuels, which may be as high as 4, 0.05 and 0.01% (all are in wt%) respectively. Chlorides and sulphates enter the engine with the air; sulphur, vanadium and sodium oxidize during combustion and mostly volatile compounds such as SO_2, SO_3, NaOH, NaO, Na_2O, VO $(OH)_3$, V_2O_5, and V_2O_4 are formed. These compounds condense at 500 to 900^0 C and build up deposits depending on the fuel. Na_2SO_4 is the main component of deposits in engines running on high sulphur and low vanadium fuel. Table I summarizes the main contaminants likely to be formed in working environment for three types of gas turbine engines i.e. aero, marine and industrial gas turbines [8]. Table II provides the composition of slag actually formed on first stage vanes in gas fired stationary gas turbine during operation [9]. Table III provides the relevance and importance of hot corrosion in various energy fields and the associated contaminants, which are responsible for degradation of components during the service [10].

Table 1. Contaminants in gas turbine fuel and air [8]

Engine application	Contaminants
Aero	Na, Cl, S, Ca (All low)
Marine	Na, Cl, S, Mg (All high)
Industrial	Na, V, S, Pb, Cl

Table 2. Composition of slag (all are in wt%) [9]

Na_2SO_4	4.3
$CaSO_4 2H_2O$	22.7
Fe_2O_3	22.3
$ZnSO_4 H_2O$	20.6
K_2SO_4	10.4
MgO	2.8
Al_2O_3	6.5
SiO_2	10.4

Temperature : 850^0 C

Atmosphere : Air containing 0.015 vol-%SO_2 -0.015 vol-%SO_3

Flow rate: 60 l h^{-1}

Table 3. Relevance and importance of hot corrosion in energy systems [10]

Energy systems	Mixed oxidant reactions	Hot corrosion
Batteries and Fuel cells	SO_2 ,SO_3 ,O_2, H_2, H_2S and H_2O attack of alloys	Fused halides and carbonate cells
Coal conversion and combustion systems	CO,CO_2 , H_2 , H_2S and H_2O attack of alloys	Slag films, residue films, fly ash films, salt films in MHD*, salt film assisted coal combustion
Solar energy and energy storage	-	Salt spillage or leakage from thermal storage tanks
Nuclear Energy	H-O-C in HTGR Steam reactions with Zircaloy	Fission product, salt condensation on cladding
Gas Turbines	SO_2, SO_3 and O_2 attack of alloys	Na_2SO_4 , NaV_xO_y , NaCl attack of alloys
Gas and Oil recovery and Magma Energy	H_2, H_2S and H_2O attack of alloys	-

* Magneto Hydrodynamic

2.1. Types

Two different forms of attack have been identified, types II hot corrosion, and I, which occur over different temperature ranges; 600-750^0 C and 800-950^0 C respectively. Type I results from a fluxing process where modification of the sodium sulphate deposit chemistry permits ingress of sulphur into the underlying metal; this produces localized depletion of protective elements and progressive internal attack occurs. The net process of type I hot corrosion produces a characteristic of attack which includes a porous oxide scale, an irregular metal / scale interface and internal attack with preceding metal sulphides.

On the other hand, type II hot corrosion requires sodium sulphate and sufficient sulphur trioxide to maintain low melting deposit, which readily fluxes the surface oxide. Thermodynamics favour sulphur trioxide formation in the lower temperature range. The associated attack is normally localized, producing pits with a lamellar scale rich in sulphur through progressive fluxing action of the deposits; sulphur generally does not enter the alloy to form internal sulphide with this type of attack [6, 11].

2.2. Mechanism

Many mechanisms have been proposed for hot corrosion. All the mechanisms involve deterioration of the reaction product barrier that forms on the alloys when the deposits are not present. Basically, the hot corrosion process proceeds in three stages (Fig.3);

a. An incubation stage during which the reaction proceeds at a rate essentially similar to that of normal oxidation

b. An initiation step during which the corrosion is accelerated

c. A propagation stage during which rapid corrosion takes place and

d. Ultimate failure of a component

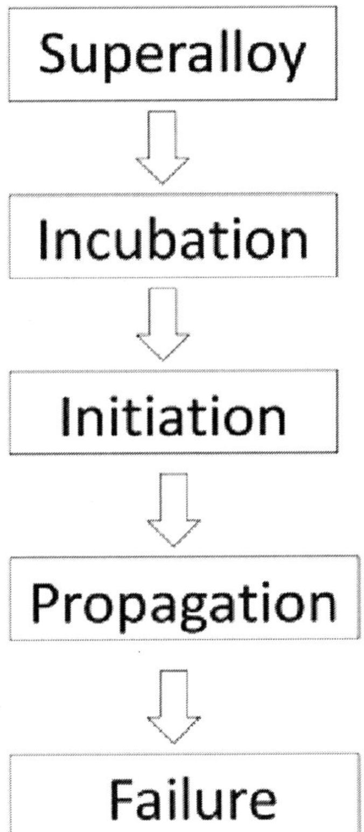

Figure 3. Various stages of hot corrosion processes, which lead to catastrophic failures of superalloy components

2.2.1. Incubation Period

During this period, the alloy undergoes normal oxidation similar to that observed in the absence of salt deposit. Initially, the oxides of most of the alloying elements are formed as given below:

$$Ni + 1/2\ O_2 = NiO$$
$$2\ Cr + 3/2\ O_2 = Cr_2O_3$$
$$2\ Al + 3/2\ O_2 = Al_2O_3$$

Due to the reaction with oxygen in the initial stages, rapid weight gain of the alloy takes place. At the end of incubation stage, thermodynamically stable oxides such as Cr_2O_3, Al_2O_3 (depending on the composition) are formed as a dense oxide scale on the surface of alloys. This oxide scale acts as a diffusion barrier for the ingress of deleterious species such as oxygen and sulphur.

2.2.2. Initiation stage

Cracking or spalling of oxide scale occurs in this stage due to the stresses developed during oxide growth. Thus, fresh alloy surface, which is depleted of scale forming alloying elements, is exposed to the action of deposit as shown below:

$$Ni + SO_4^{2-} = NiO + SO_2 + O^{2-}$$

SO_2 will be in the form of dissolved gas.

2.2.3. Propagation stage

The propagation stage during hot corrosion is substantially different from the behaviour of the alloy in the absence of a deposit. It is accompanied by a very severe attack of the alloy as a result of fluxing of oxides as shown below:

$$2\ NiO + 2\ O^{2-} = 2\ NiO_2^{2-}\ (\text{nickelate})$$

$$Al_2O_3 + O^{2-} = 2\ AlO_2^{-}\ (\text{aluminate})$$

$$Cr_2O_3 + 2\ O^{2-} + 3/2\ O_2 = 2\ CrO_4^{2-}\ (\text{chromate})$$

The dissolved sulphur dioxide formed during initiation period can react with the alloying elements to produce oxides, sulphides and sulphur.

2 Ni + SO 2 = 2 NiO + 1/2 S 2 4/3 Cr + 3/2 Ni + SO 2 = 2/3 Cr 2 O 3 + 1/2 Ni 3 S 2 2 Ni + SO2 = 2 NiO + 1/2 S24/3 Cr + 3/2 Ni + SO2 = 2/3 Cr2O3+ 1/2 Ni3S2

The relationship between the initiation and propagation stages and the factors that are important in preconditioning the alloy during the initiation stage are explained by Pettit and Meurer [12].

If the deposits contain chloride ions, it can selectively remove certain elements such as chromium or aluminium from the alloy. This process involves the formation of highly volatile gaseous chlorides inside the pores of the alloy and thereby metal chlorides diffuse out from the alloy. As a result, mechanical properties of the alloy reduce significantly as the cracks are formed on the alloy surface. The metallic components of these chlorides convert to oxides eventually but these oxides form as particles and not as continuous layers. Hence, severe attack of the alloy can take place.

The mechanisms have been explained based on acidic and basic fluxing of protective oxide scales depending upon the conditions. The protective oxide scale, for example Cr_2O_3 is fluxed either as acidic solutes such as $Cr_2(SO_4)_3$ or CrS or as basic solutes such as Na_2CrO_4 and $NaCrO_2$. As a result, the oxide scale can become non-protective. Such non-protective scales can be formed by dissolution of the oxide near the alloy surface and re-precipitation as discontinuous particles from the molten deposit. It is important to emphasise that there are other means by which an oxide scale can become non-protective via a fluxing process. For example, the fluxing reaction may cause a certain element to be removed from the alloy, which in turn gives rise to a non-protective scale on the surface of the alloy. Rapp and his group [13-14] carried out extensive studies pertaining to hot corrosion fluxing mechanisms, different oxide solubilities and electrochemical evaluation methods for hot corrosion resistance. Kinetic studies of various materials at high temperatures under hot corrosion conditions were reported by the earlier workers [15-16]. Natesan et al [17-18] have studied the high temperature corrosion of Ni- based superalloys in coal conversion environments.

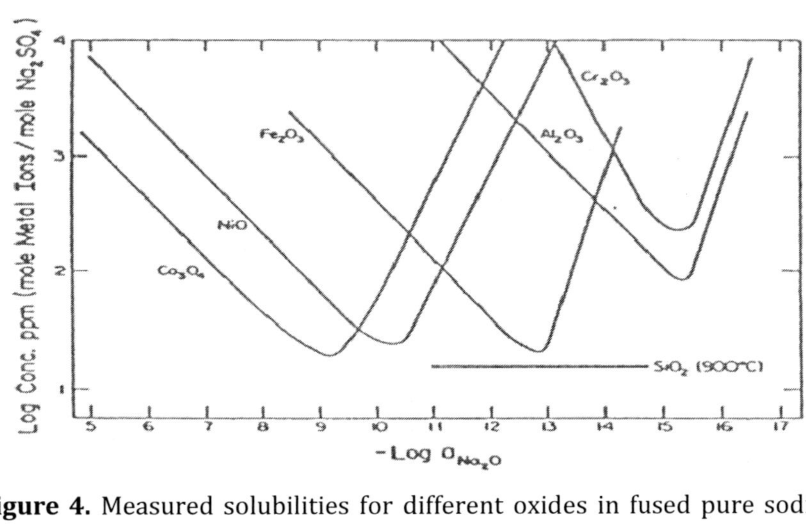

Figure 4. Measured solubilities for different oxides in fused pure sodium sulphate at 1200 K [13]

As fluxing of the protective scale is important in hot corrosion attack, the selection of alloys or coatings for hot corrosion resistance should be based on the solubility of protective oxides. For a given deposit and environmental conditions, the most favoured oxide is the one that has the least solubility and can form a protective scale in the presence of salt film. Fig.4 presents the measured solubilities of various oxides in fused Na_2SO_4 at 1200 K [13]. The difference of six orders of magnitude in basicity between minima for the most basic Co_3O_4 to the acidic SiO_2, Cr_2O_3 and Al_2O_3 is striking and is consistent with important known alloy systems and coatings under hot corrosion conditions. Cobalt based alloys and coatings are more vulnerable to acidic fluxing than Ni-based alloys and coatings. Cr_2O_3 is resistant to acidic fluxing because the minimum in its solubility curve corresponds approximately to the acidity of gas turbine environments.

It is very essential to assess the materials / coatings in the laboratory under simulated engine conditions for testing their performance, since it is not feasible to test each material directly in the engine. A satisfactory test should yield a prediction of service performance, with an estimation of component life as a desirable objective. The next section covers different techniques, which are used for evaluation of various materials / coatings for hot corrosion resistance.

2.3. Assessment techniques

The following laboratory hot corrosion techniques are in use for assessing a variety of materials /coatings for gas turbine engine applications:

i. Burner rig test

ii. Furnace test

iii. Crucible test

iv. Thermo gravimetric test

v. Electrochemical methods

2.3.1. Burner Rig test

This test simulates the operating conditions in gas turbine engines, the gas composition, pressure, velocity and temperature. In this test, rigs are composed of a combustion chamber from a small turbine, supplying air from compressors and burning fuel in the usual way. Salt can be sprayed into the combustion chamber. The hot exhaust gases are brought into the sample chamber, where several specimen coupons are placed and rotated in the gas stream. Here, the test variables (such as gas pressure, velocity, sample temperature, salt concentration and fuel-to-air ratio) can be selected to simulate aero, marine and land based gas turbine engine's operation. This test was used extensively for evaluation of coated and uncoated materials for hot corrosion resistance [19-22]. It was also established that the materials tested showed generally similar behaviour during service. Therefore, this test is more appropriate for testing the materials as well as coatings to get comparable results with the service conditions of gas turbine engines.

2.3.2. Furnace test

In this test, a furnace having two zones, whose temperature can be controlled independently is employed. The specimen is placed in one zone and a crucible containing the test salt is placed in the other. Using a carrier gas, the vaporized salt is carried from one zone into the other where it deposits on the specimen. In this manner, the deposition of salt per unit time can be strictly controlled. In modified method known as 'Dean Test', the corrosion conditions can be controlled from very mild to very severe, by simply altering the temperature differential within the two zones. This test was used mainly to study kinetics of various materials in different environments.

2.3.3. Crucible test

This method involves direct immersion of the sample to half its length in molten salt in air atmosphere and measurement of weight change to monitor the reaction rate. Gas may be bubbled through the melt or passed over it to provide oxidizing or reducing environments. This test was used by Simons et al [23] for the first time followed by other researchers for evaluation of a number of bare and coated materials for hot corrosion resistance. The corrosion test conditions are severer than that in normal operating condition of the engines, but it is a simple and helpful test for screening of different materials and coatings [24-34].

2.3.4. Thermo gravimetric test

It is also called "Salt coated test". In this test, a coupon of the material is coated with saturated salt solution, then dried, weighed and placed in a heated furnace. The coupon weight is continuously recorded. In most cases, the weight increases as oxidation proceeds, as the vaporization rate is small compared with the oxidation rate. This method was adopted by many researchers for determining the hot corrosion kinetics because the experimental parameters such as gas composition, temperature and salt loading can be controlled very easily and precisely.

2.3.5. Electrochemical methods

This method involves measurement of corrosion current as a result of oxidation-reduction reaction continuously and provides instantaneous corrosion rates. The net reaction is the oxidation of metal to form oxide scales and the reduction of SO_4^{2-} to a lower oxidation state. Generally, corrosion is accelerated only when the salt film is molten such that the material and the gas are physically separated by the ionically conducting fused salt. Fused sodium sulphate is a dominant ionic (Na^+) conductor like in normal solutions. Thus, the reaction must involve electrochemical steps and hence this method is most ideal for assessing materials / coatings for hot corrosion resistance.

Table 4. Comparison of hot corrosion evaluation techniques

Technique	Advantages	Disadvanges / limitations
Burner Rig Test	Simulates gas composition, pressure, velocity and temperature of gas turbine engines	Complex process and requires long running time. Difficult to control all the parameters accurately
Furnace Test	Corrosion conditions can be controlled depending upon the requirements i.e. severe or mild corrosion	Difficult to maintain salt deposition rate for longer times
Crucible Test	Most simple and highly useful for preliminary screening of the materials and coatings	Corrosion is severe for alloys or coatings having low or intermediate resistance to hot corrosion
Thermo-gravimetric Test	Precise weight gain measurement possible under different test conditions i.e. gas composition, temperature, salt composition etc.	Not useful for predicting the life of materials and coatings
Electrochemical Test	Fast and useful for ranking the materials and coatings based on oxidation-reduction Phenomenon	Yet to be established fully

The electrochemical polarization of alloys in molten salts can cause corrosion attack which is qualitatively similar to that found for a thin salt film in a combustion- product environment. In addition, the 600 hours burner rig test results are well comparable with 4 hour electrochemical test method. It proves that electrochemical techniques are extremely helpful in evaluating the materials / coatings at a faster rate compared to other tests developed so far and hence this method was used extensively by the researchers for the above purpose as well as to study reaction mechanisms and on line monitoring of corrosion rates in the actual plants [35-37]. The advantages and disadvantages of each technique are presented in Table IV. It is important to mention here that no established ISO standard testing procedure is available till now for evaluation of materials and coatings for their hot corrosion resistance and hence the need to establish such standards in near future. In fact, it is an important issue that needs special attention by all researchers in the field.

2.4. Influence of superalloy composition

Different superalloys of varied categories namely forged alloys such as Nimonic 75, Nimonic 105, Inconel 718 etc., conventional casting (CC) alloys like Inconel 713, Inconel 100 etc., and directionally solidified alloys such as CM 247 LC, MAR-M200, MAR-M247 etc. were reported for their hot corrosion resistance in different environments. Single crystal superalloys of different generations such as first generation namely CMSX-2, TMS-12, TMS-26, PWA1480, Rene N4 etc., second generation with up to 3 mass % rhenium (CMSX-4, Rene N5, TMS 82+ etc.), third generation with up to 6 mass % rhenium (CMSX-10, TMS-75, TMS 80+ etc., and fourth generation with rhenium and ruthenium (TMS-138 etc.) have been developed recently. The 3rd and 4th generation superalloys contain large amount of rhenium (Re) as mentioned above and less amount of chromium, exhibits good hot corrosion resistance but susceptible to oxidation. As one of the directionally solidified (DS) superalloys, CM 247 LC alloy exhibits comparable properties as the single crystal alloys. Table V presents different superalloys for which hot corrosion characteristics have been reported in the literature [6, 32, 38-44]. Advancements in superalloys as well as processing techniques have permitted the hot gas path components now to operate in modern gas turbine firing at increased temperatures for significantly increased duration (many thousands of hours) under severe conditions of centrifugal, thermal and vibratory stresses.

Few superalloys have been designed and developed based on phase stability and utilizing predictive techniques by balancing the levels of critical elements (Cr, Mo, Co, Al, W and Ta) in order to maintain good hot corrosion resistance at higher strength levels. One among them is IN 738 superalloy. Therefore, IN 738 is notable as being one of a very small class of modern superalloys that has an outstanding combination of elevated temperature strength and hot corrosion resistance. The balance of these two properties was optimal for heavy-duty gas turbine applications. It was significantly designed for application in land-based gas turbine, as opposed to aircraft use. This alloy was first developed by the international nickel company and subsequently modified its chemistry to improve its castability. This work enabled the successful application of IN 738 over the past 20 years in General Electric gas turbines. Indeed this alloy is now used throughout the entire heavy-duty gas turbine industry with platinum aluminide coatings. Developmental efforts were focused on single processing and advanced DS alloy development. Single crystal airfoils offer the potential to further improve component high temperature materials strength and by control crystal orientation, can provide an optimum balance of properties. In single crystal material, all grain boundaries are eliminated from the material structure and a single crystal with controlled orientation is produced in an airfoil shape. By eliminating all grain boundaries and the associated grain boundary strengthening additives, a substantial increase in the melting point of the alloy can be achieved, thus providing a corresponding increase in high temperature strength. The transverse creep and fatigue strength is increased compared to equiaxed or DS structures. The advantages of single crystal alloys compared to equiaxed and DS alloys in low cycle fatigue (LCF) are shown in Fig.5. The single crystal alloys (2nd generation) have been in use in gas turbine engines since 1995. Together with improved protective coatings, the new materials like third and fourth generation alloys will provide enhanced growth capability for gas turbine engines in the years to come.

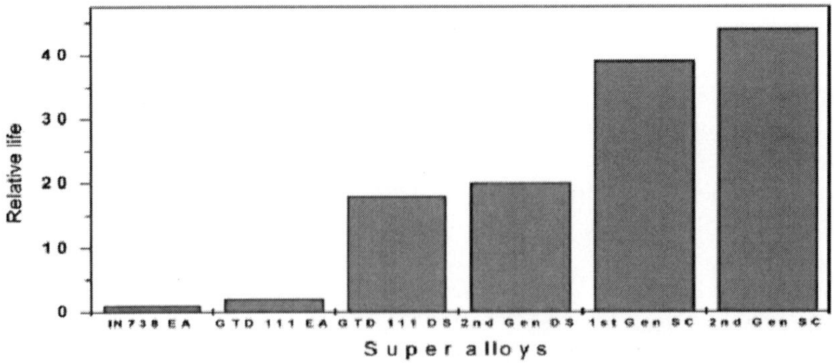

Figure 5. Low cycle fatigue (LCF) of various superalloys

Single crystal superalloys containing rhenium and iridium with low chromium, TMS-75 and TMS-75 + Ir respectively, was observed to exhibit balanced intermediate and high temperature creep, strengths with good hot corrosion resistance based on short term laboratory studies (20 hours only) in 75% Na_2SO_4 + 25% NaCl (Fig.6) [38]. However, further research is needed on these materials on long term basis with aggressive species like NaCl and V_2O_5 in order to establish their characteristics.

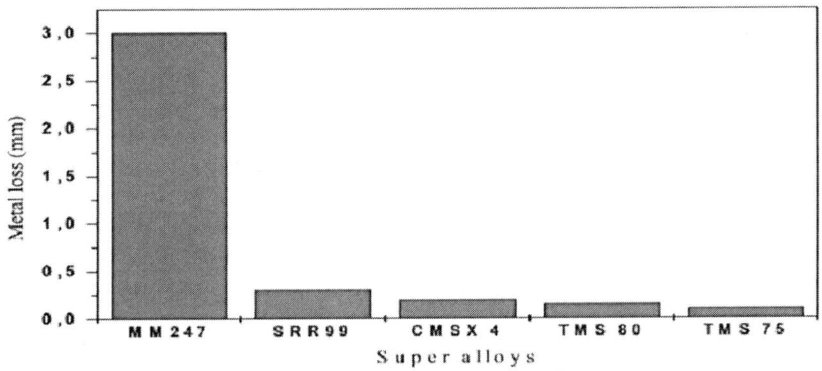

Figure 6. Hot corrosion resistance of different superalloys in 75% Na_2SO_4 + 25% NaCl at 900^0 C for 20 hours [38]

Fig.7 shows the hot corrosion behavior of few superalloys like IN 792, CMSX-4 and DMS-4 under type I conditions. The hot corrosion resistance of CM 247 LC and Rene 80 Superalloys under type II and type I conditions in chloride and vanadium environments are also presented in Fig.7. Appreciable corrosion was observed for all the superalloys. It indicates that hot corrosion plays a significant role in causing faster degradation, thereby reducing the superalloy life considerably. Among the superalloys, CM 247 LC was corroded severely indicating that the superalloy is highly susceptible to hot corrosion. The appreciable corrosion attack of CM 247LC superalloy was clearly evidenced by observing big cracks, broken samples (Fig.7) and large corrosion affected zone (due to appreciable diffusion of corrosive elements present in the environment). Sulphur diffusion and formation of metal sulphides preferentially chromium and nickel sulphides was reported to be the influential factor [6]. When sulphide phases are formed in superalloys, Ni- based alloys are inferior to cobalt and iron based alloys, which are especially effective in destroying the corrosion resistance of alloys [26, 36-42]. In fact, the alloying elements play a significant role and decide the life of superalloys under hot corrosion conditions.

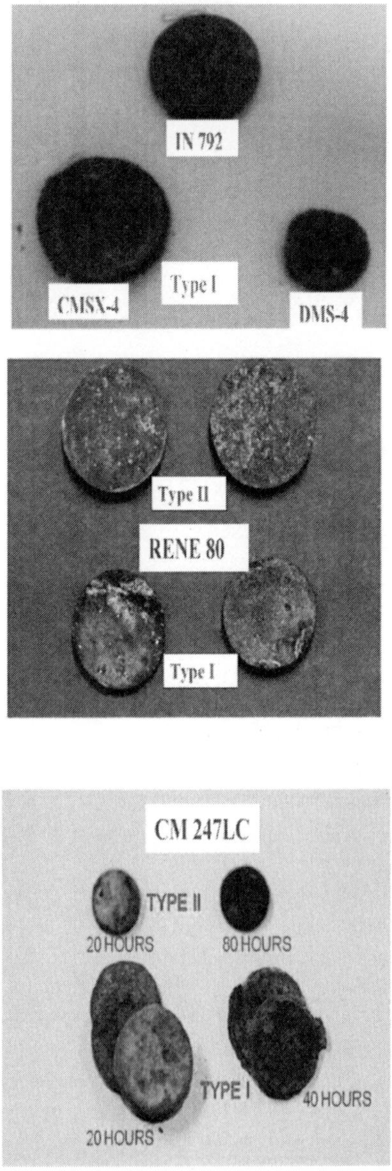

Figure 7. As hot corroded superalloys in various environmental conditions

Table 5. Typical superalloys characterised for their hot corrosion resistance [6, 32, 38-44]

Alloy	Al	Ti	Cr	Co	Ni	W	Ta	Re	Mo	Hf	Ir	Others
NIMONIC 75	-	0.4	20	-	Bal	-	-	-	-	-	-	Fe, C
NIMONIC-105	4.7	1.3	15	20	Bal	-	-	-	5.0	-	-	C
INCONEL 100	5.5	5.0	10	15	Bal	-		-	3.0	-	-	Zr,C,B,V
INCONEL 713	6.0	0.8	12.5	-	Bal	-	1.75	-	4.2	-	-	Zr,C,B,Nb
MAR M200	5.0	2.0	9.0	10	Bal	12.5	-	-	-	-	-	Zr,B,C,Nb,Fe
IN 738	3.4	3.4	16	8.3	Bal	2.6	1.75	-	1.75	-	-	Cb,Fe,C
CMSX-4	5.3	1.0	6.5	10.4	Bal	6.1	4.9	0.8	-	-	-	-
MAR M-247	5.5	1.0	8.0	10.0	Bal	10.0	3.0	-	0.6	1.5	-	B,Zr,C
MAR M-509	-	0.2	23.5	55	10	7	3.5	-	-	-	-	Zr,C
CM 247LC	5.6	0.7	8.1	9.2	Bal	8.5	3.2	-	0.5	1.4	-	B,Zr,C
INCONEL 718	-	-	18	-	Bal	-	-	-	3.0	-	-	Fe,Cb-
CMSX-10	5.7	0.2	2.0	3.0	Bal	5.0	8.0	6.0	0.4	0.03	-	-
TMS-75	6.0	-	3.0	12.0	Bal	6.0	6.0	5.0	2.0	0.1	-	-
TMS-80	5.8	-	2.9	11.6	Bal	5.8	5.8	4.9	1.9	0.1	3.0	-

Some Ni-based superalloys were designed with the aid of d-electrons concept for industrial turbines. They developed based on the molecular orbital calculations of the electronic structures of Ni-alloys i.e. based on two electronic parameters such as bond order between an alloying element and nickel atoms Bo, and d-orbital energy level of alloying element Md. Some single crystal superalloys were designed based on this concept. The experimental data on the designed single crystal superalloys was reported to be satisfactory [38-39]. However, the need is essential in order to establish the concept for development of hot corrosion resistant Ni-based superalloys as the superalloys should satisfy both high temperature strength and high temperature corrosion resistance, which is a highly difficult task. It is due to fact that same alloying elements never help to improve both high temperature strength and corrosion resistance simultaneously as mentioned earlier.

From the metallurgical point of view, it is known that high temperature strength can be obtained by maintaining certain phases that are responsible for high temperature strength. Since the main motive for metallurgists is to improve the mechanical strength of an alloy at elevated temperatures, the addition of certain alloying elements is essential in order to form γ' and solid solution strengtheners. Among the alloying elements, the significant reduction of chromium content and addition of certain elements like tungsten, molybdenum, vanadium etc. makes the superalloys highly susceptible to hot corrosion [6, 26, 41]. Tungsten (W) and Mo dissolve in γ-phase of the matrix and strengthen the alloy. These elements are particularly effective for improving long-term strength of the alloy. However, excessive addition causes the formation of harmful sigma phase and affects the strength of the alloy. Suitable contents for obtaining good strength are 2 to 15% of W and 6% or less of Mo. The addition of titanium and tantalum produces beneficial effects while small additions of manganese, silicon, boron and zirconium do not significantly influence the hot corrosion resistance of superalloys. Carbon addition is detrimental as the carbide phases provide sites for initiation of hot corrosion [26]. With a large content of chromium and small amount of titanium, hot corrosion resistance of Nimonic-75 is very good [32]. The addition of molybdenum and iron made the Inconel 718 less corrosion resistant [6]. The addition of 9.5% tungsten and minor additions of other elements rendered CM 247LC highly susceptible to hot corrosion [6]. The tungsten present in the alloy forms acidic tungsten oxide (WO_3), due to which fluxing of protective oxide scale takes place easily. It is important to mention that this type of acidic fluxing is self-sustaining since WO_3 forms continuously that leads to severe degradation of superalloys. The degradation mechanism comprises two steps as follows:

a. The tungsten present in the superalloys reacts with the oxide ions present in the environment and forms tungsten ion

$$WO_3 + O^{2-} = WO_4^{\ 2-}$$

b. As a result, the oxide ion activity of the environment decreases to a level where acidic fluxing reaction with the protective alumina can occur

$$Al_2O_3 = Al^{3+} + O^{2-}$$

A similar reaction mechanism occurs if the superalloys contain other refractory elements like vanadium and molybdenum [26, 40].

Cobalt (Co) and chromium (Cr) contents affect the hot corrosion resistance of superalloys. If Co is added excessively, it promotes precipitation of harmful inter-metallic compounds that deteriorates high temperature strength. An appropriate content of Co is 10.5% or less. Cr improves hot corrosion resistance. However, if it is added excessively, it causes precipitation of harmful sigma-phase and coarsening of carbides, thereby deteriorating high temperature strength. An appropriate additive amount of Cr is in a range of 5 to 14%. Aluminium and titanium contribute to improvement of high temperature strength by forming γ'-phase i.e. $Ni_3(Al,Ti)$, which is a strengthening factor of the Ni-based alloys. They improve hot corrosion resistance. Appropriate amounts of Al and Ti are in ranges of 4 to 7% and 0.5 to 5% respectively.

Nb, Ta and Hf dissolve in γ'-phase, which is a strengthening factor and improve high temperature strength of the superalloy. However, if they are added excessively, they segregate at grain boundaries and reduce the strength of the alloy. Appropriate additive contents of Nb, Ta and Hf are, respectively, 3 % or less, 12% or less and 2% or less and preferably 0.2 to 3% of Nb, 3 to 4% of Ta and 0.5 to 1.0% of Hf. Zr and B strengthen grain boundaries and improve high temperature strength of the alloy. Since the addition of these elements are in the low range and there is no much effect on hot corrosion resistance of the superalloys.

The common alloying element for the new generation superalloys is rhenium [41-44]. In rhenium containing superalloys, the Re atoms dissolve preferentially in γ matrix, retards the over-ageing of γ' dispersed phase thereby increase γ-γ' misfit. In addition, Re atoms form small domains (clusters) of approximately 1nm diameter, act as barriers to the dislocation movements. The formed clusters improve the creep resistance of the superalloys. Kobayashi et al have reported that Re also helps in improving the hot corrosion resistance though significantly low chromium is present in superalloys [38]. However, the exact mechanism based on which the hot corrosion resistance improved was not reported. It was also reported that rhenium containing superalloys are susceptible to oxidation. It indicates that the oxide scale that forms upon exposure is not protective. It should not form either alumina or chromia scale due to significantly low contents of aluminium and chromium. Extensive studies have been carried out recently with and without Re containing superalloys in order to understand the role of rhenium and confirmed that Re containing superalloys are highly susceptible to high temperature oxidation as well as both types of hot corrosion [41, 45-47].

2.4.1. Degradation mechanisms

The mechanisms proposed earlier are sulphides formation and subsequent transformation to corresponding alloy element oxides. Later, it was reported that degradation of superalloy take place due to fluxing processes. Fluxing processes are divided into two types such as acidic and basic fluxing process. The nature of fluxing process depends on the alloying elements and their concentration in the superalloys. It also depends on the environment to which the superalloys are exposed. In the fluxing process, the corresponding oxides dissolve in the components of environment and cause to degrade the superalloys at a much faster rate.

Recently, an electrochemical mechanism was proposed and experimentally proved that all the superalloys, irrespective of their alloying elements and concentration and any type of environments to which they are exposed, degrade due to electrochemical reactions that take place during hot corrosion process. Hence, the electrochemical techniques are more suitable for assessing newly developed superalloys and their ranking [6].

3. COATINGS

There are a number of cases reported in the literature where the gas turbine blades suffered severe corrosion due to which failures took place [2-5]. Failure investigations confirmed that it was due to hot corrosion, in which extensive penetration of sulphur took place into the material leading to the formation of metal sulphides that in turn led to reduction in mechanical properties of the materials resulting in catastrophic failures (Figs. 8, 9, 10). From literature it is also understood that it is not possible to develop an alloy having both high temperature strength and hot corrosion resistance simultaneously since alloying elements often have opposing effects on the two properties. Particularly, tungsten, molybdenum and vanadium additions to superalloys are helpful in improving the high temperature strength, but their presence make the superalloys highly susceptible to hot corrosion as mentioned above. High performance coatings that can withstand environmental degradation problems such as oxidation and hot corrosion become important. Therefore, it is mandatory to prevent hot corrosion by using appropriate coating technology [41, 46] so as to increase the life of gas turbine components to the designed service life (Fig.11).

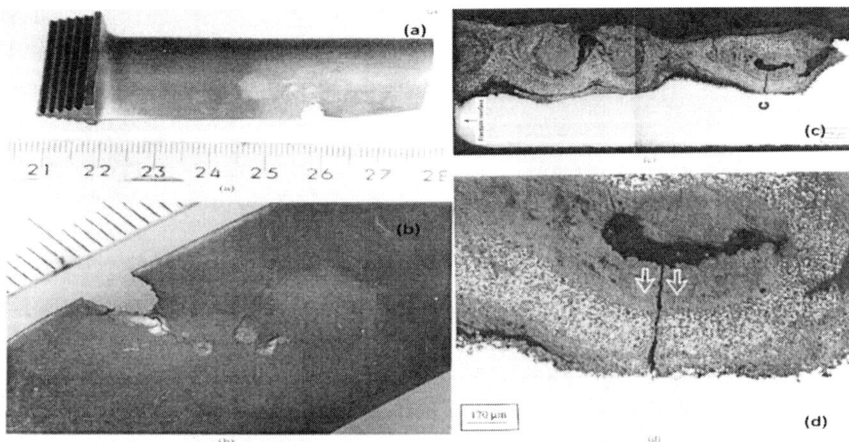

Figure 8. a) Macroscopic view of damaged blade b) a crack propagating into remainder material c) micrograph showing advanced hot corrosion attack d) secondary crack propagation [3]

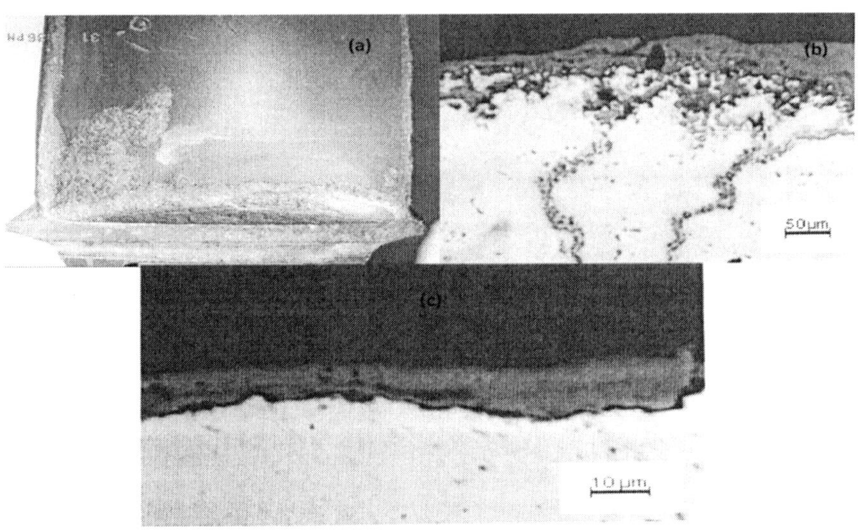

Figure 9. a) Macroscopic view of hot corrosion affected blade b) grain boundary diffusion and subscale sulphide particles indicating type I hot corrosion c) no grain boundary diffusion and subscale sulphide particles indicating type II hot corrosion [4]

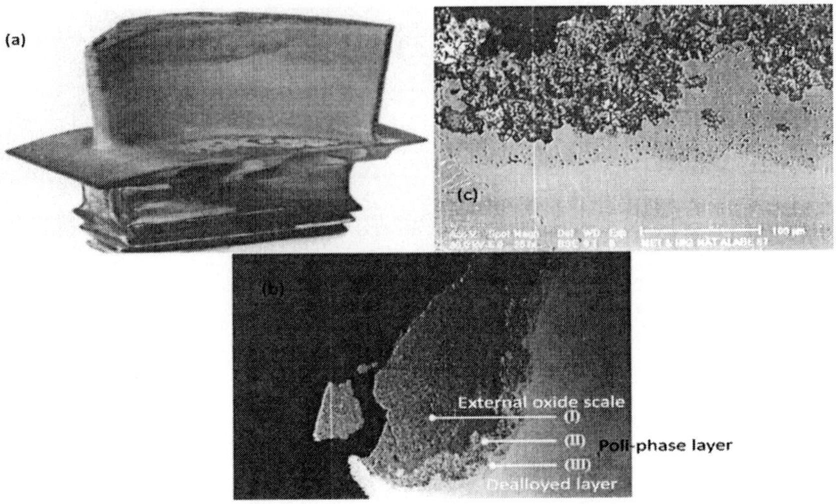

Figure 10. a) Macroscopic view of failed blade b) corroded region consisting of three layers c) internal layer of corrosion products [5]

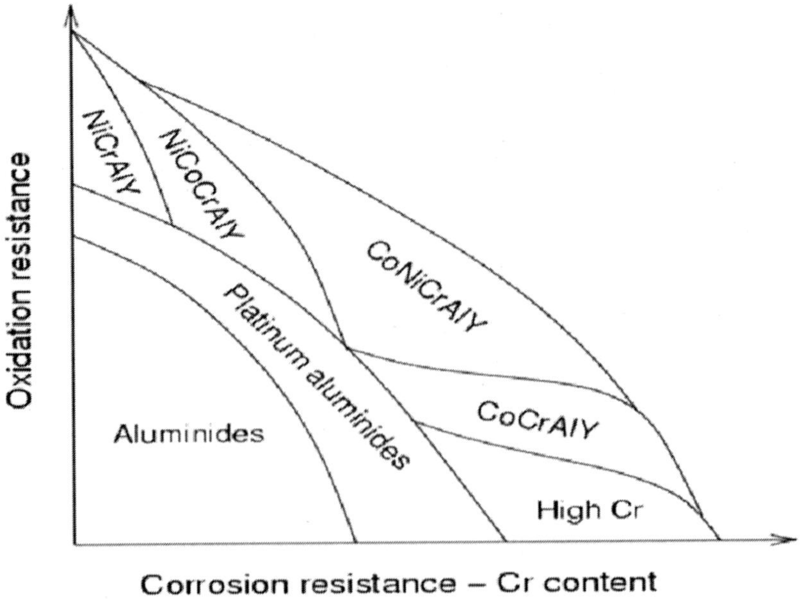

Figure 11. Oxidation and hot corrosion resistance of various coating systems [46]

The blade coatings are usually of either MCrAlY type (where M is Ni or NiCo) or aluminides. These coatings can provide protection against oxidation and hot corrosion and act as bond coatings for zirconia based thermal barrier coating (TBC) systems. In both the cases, slow growth rates and optimum adherence of the alumina scales formed on the coatings during high temperature exposure are of significant for component life. These requirements can be fulfilled only by using coatings with sufficiently high aluminium contents to ensure protective alumina scale formation and re-healing after oxide spallation / reaction with the environment. The life of a coating is mainly limited by aluminium depletion occurring upon aluminium consumption as a result of alumina scale growth and repeated spallation and re-healing of the alumina scale during oxidation process. If a point is reached where the aluminium level in a bond coating falls below the level at which protective alumina scale can not be formed preferentially, faster interaction between the corrosive species present in the environment and the non-protective oxides of other constituents of the bond coating occurs and thereby affects the coating life considerably under hot corrosion conditions. Further, the constituents of ceramic thermal barrier coatings react easily with the corrosive species and shorten the coating life significantly. These aspects are covered in detail in subsequent sections.

Another important aspect is the selection of suitable surface engineering technique by which the coatings are applied since the life of a coating depends not only on coating composition but also on the surface engineering technique employed for coating. Therefore, selection of appropriate surface engineering technique as well as suitable coating composition becomes a challenging task.

3.1. Technologies

Surface engineering technique, which modifies properties of the component's surface to enhance their performance should not only be effective but also economical. The following are the most common surface engineering techniques available at present:

a. Electro-deposition

b. Diffusion coating processes

c. Thermal spray techniques

d. Ion implantation

e. Hardening & cladding

f. Selective surface hardening by transformation of phase and

g. vapour deposition.

Electro-deposition, diffusion and thermal spraying techniques are widely in use to improve the surface of components for hot corrosion resistance. Particularly, thermal spray and vapour deposition techniques are the most efficient techniques for prolonging the life of components significantly.

3.2. Thermal spraying processes:

The coatings applied by these processes have the advantage over other techniques i.e. the coating composition can be tailored to suit the specific requirements. Thermal spraying processes can broadly be divided into following techniques:

i. Flame Spraying

ii. Arc Spraying

iii. High Velocity Oxyfuel Spraying

iv. Detonation Gun

v. Air Plasma Spraying

vi. Vacuum Plasma Spraying

3.2.1.1. Flame Spraying

In this process, an oxygen-acetylene mixture is passed through a nozzle and ignited to form a combustion flame. Coating powder or wire is fed into the flame, accelerated and projected onto the substrate to form deposit. The combustion flame temperature is limited to 3000^0 C and gas particle velocities are relatively low.

3.2.1.2. Arc spraying

This process involves the production of molten particles at the tips of two consumable wires via resistance heating. The material is subsequently atomized and projected onto the substrate by a compressed air jet. The process is limited to the spraying of conductive wires, relatively cheap and can achieve high deposition rates.

3.2.1.3. High Velocity Oxy-fuel Spraying (HVOF)

This process involves combustion of a fuel gas with oxygen at high pressure. This creates high velocity flame propelling powdered coating materials onto a substrate. This technique can produce high quality coatings for gas turbine applications.

3.2.1.4. Detonation gun spraying

D-Gun and Super D-Gun are developed by Union Carbide, use the energy released by a controlled series of oxygen-acetylene to heat and accelerate the coating powder to high velocities and to propel the powder onto the substrate. The resulting coatings are highly dense and of high quality with a low content of oxide and unprocessed particles.

3.2.1.5. Air Plasma Spraying (APS)

In air plasma spraying, a DC arc struck between a central inert cathode and an annular copper anode is used. An inert gas is fed into the arc to form high temperature plasma. Powder is fed into the plasma and is ejected at a high velocity towards the substrate.

3.2.1.6. Vacuum Plasma Spraying (VPS)

This process has several advantages over APS. The problems associated with air contamination of the plasma jet gets eliminated, the plasma jet is longer than in air and can achieve 400-600 ms^{-1} particle velocities due to which high purity and dense deposits are obtained. A further advantage is coating adhesion. Table VI illustrate the characteristics of various thermal spraying processes [47].

Table 6. A comparison of thermal spraying processes [47]

Process	Gas temperature^0	Particle Velocity m/s	Adhesion Mpa	Oxide content %	Porosity %	Deposition Rate Kg/Hour	Typical Deposit thickness mm
Flame	3000	40	8	10-15	10-15	2-6	0.1-15
Arc wire	N/A	100	12	10-20	10	12	0.1-50
HVOF	3000	800	>70	1-5	1-2	2-4	0.1-2
Detonati-on gun	4000	800	>70	1-5	1-2	0.5	0.05-0.3
APS	12000	200-400	4 to >70	1-3	1-8	4-9	0.1-1
VPS	12000	400-600	>70	ppm	<0.5	4-9	0.1-1

Kuroda et al were developed a novel coating technique namely "warm spray" process recently which incorporates the advantages of both HVOF and cold spray [48]. Further developments may enable them to develop coatings with enhanced hot corrosion resistance.

3.3. Electron Beam Physical Vapour Deposition (EB-PVD) processes

The electron beam-PVD process has made the most significant impact over the PVD processes in respect of coating rotating blades. The microstructure of the coating is controlled very closely in this process and coating process is slow compared to conventional PVDs. Thermal barrier oxide coatings obtained with this process has a unique columnar structure with high in-plane compliance, which tends to have higher thermal conductivity and has shown to survive longer thermal cycle life as compared to plasma sprayed TBC. Table VII provides the comparison of different processes and their limitations [47].

Table 7. Comparison of thermal spray and EB-PVD processes [47]

Coating type	Advantages	Disadvantages /limitations
Diffusion (e.g. Aluminides)	Simple to produce, proven ability <900⁰ C, inexpensive	Thickness limited to 50 μm. Brittle at <750⁰ C and degrade by inter-diffusion. Available compositions are limited.
Plasma overlays	High rates of coating and wide range of compositions are available. Thickness not process limited (100 μm)	Rough surface and single line of sight coating than EB coatings. Thin multi-layers more difficult than EB. About 15% porosity and degrade by inter-diffusion.
EB-PVD overlays	Good control of microstructure, low contaminant level and control the composition within limits. Layered and graded coatings easily achieved and multi component processing is possible.	Low rates of deposition and expensive process. Degrade by inter-diffusion.
Plasma Spray Thermal Barrier	High rate of coating and wide range of compositions are available. Layered and graded coatings are possible. Porosity is an advantage and not limited by inter-diffusion.	Single line of sight coating and difficult to modify microstructure. High stresses generated in coatings.
EB-PVD Thermal Barrier	Good control of microstructure and low contamination level. Layered and graded coatings are possible and not limited by inter-diffusion.	Low rate of deposition and multiple component processing is possible. Expensive process and generates high stresses but relieved by structure control.

3.4. Metallic coatings

The present coatings are broadly divided into two types, diffusion and overlay.

3.4.1. Diffusion coatings

In diffusion coating application process, aluminium is made to react at the surface of substrate, forming a layer of mono-aluminide. For coatings applied over Ni-based superalloys, nickel aluminide is the resulting species. This type of coatings and surface modification is one of the most widely used for tailoring the surface properties of components. Pack cementation is well-established chemical vapour deposition technique, which has been in use for nearly three decades. It is relatively a simple technique, which consists of immersing the components to be coated in a powder mixture placed in a sealed or semi-sealed retort. The entire apparatus is placed inside a furnace and heated in a protective atmosphere to a high temperature for a sufficient period to form a coating. Most blade and vane coatings were applied by pack cementation since 1970 and subsequently chemical vapour deposition (CVD) process. The surfaces of internal cooling passages are coated more effectively by forced flow gas phase aluminizing and by vacuum pulse aluminizing. From 1970's onwards, developments in the area of diffusion coatings include modification of aluminide coatings with chromium, silicon and platinum. In 1990's aluminide coatings were recognised as useful bond coatings for thermal barrier coatings (TBCs). Advanced systems started using platinum aluminides and MCrAlY coatings [24, 28, 49-50]. The most developed pack cementation processes include aluminizing, chromizing and siliconising. Under this category, two types of processes can be applied; low activity, where the aluminide coating is formed by outward diffusion of nickel from the base metal, or high activity, where the aluminide coating is formed by inward diffusion of aluminium and subsequent heat treatment. High activity process is most suitable for gas turbine applications to combat oxidation but it is susceptible to hot corrosion.

The most important improvement in the diffusion coatings has been the incorporation of platinum in aluminide coatings. This process involves the deposition of platinum by electrochemical method followed by aluminizing at the suitable temperature for the required period. These coatings became the accepted standard for turbine hot components to combat oxidation as well as hot corrosion. The principal reason for superior performance of these coatings is that the coefficient of thermal expansion of the coating is lower compared to that of underlying superalloy. In addition, platinum enhances the activity of aluminium and allows the formation of continuous alumina scale during high temperature exposure coupled with good adherence, thereby prolonging the life of turbine blades considerably. Failure of coatings under turbine service conditions takes place due to inter-diffusion between the coating and the substrate, with the loss of protective capability. Though diffusion coatings are well bonded to the substrate they have limited compositional flexibility and their usefulness is strongly dependent on substrate chemistry.

3.4.2. Overlay coatings

To develop coatings with compositions nominally independent of substrates and with capabilities for tailoring to a wide range of requirements for gas turbine applications, overlay coatings were emerged. Coatings of this type are generally called MCrAlY coatings and essentially comprise a mono-aluminide component contained in a more ductile matrix of a solid solution. The supply of aluminium for formation of protective alumina scale comes largely from the dispersed mono-aluminide phase during the useful life of such coatings. Overlay coatings are typically well bonded and have a wide compositional flexibility. Research and development on this type of coatings has led to a variety of compositions with improved scale adherence. The function of all coatings is to provide a surface reservoir of critical elements that will form very protective and adherent oxide layers, thus protecting the underlying base materials from oxidation as well as hot corrosion.

Low pressure plasma spraying has been mostly used to apply MCrAlY coatings to critical components such as the high pressure rotor blades because it can produce dense coatings with little oxidation and good adhesion. It has been shown that MCrAlY coatings prepared with air plasma spraying has much inferior oxidation resistance due to severe oxidation during spraying and resultant loss of aluminium during spraying. HVOF sprayed MCrAlY has been increasingly adopted due to its dense and well-controlled oxidation.

3.4.3. Advances in metallic coatings

The efficiency of a gas turbine engine (aircraft, industrial, marine) is directly proportional to firing or turbine inlet temperature. Increase in temperature is facilitated by improved structural design and airfoil-cooling technology applied to higher strength-at-temperature alloys cast by increasingly complex processes and coated with steadily improved protection systems. First stage turbine blades, the most critical components of gas turbines, made from Ni-based superalloys and augmented by coatings, have been singularly successful materials for the past several years. The turbine airfoils coatings are simple and modified diffusion aluminides, MCrAlY overlay systems and overlay thermal barrier coatings.

The oxidation resistance of alloys and coatings depends upon selective oxidation of elements such as aluminium, chromium or silicon present in the alloy / coating. Oxide scales of Al_2O_3 are more effective means for developing oxidation resistance at higher temperatures and Cr_2O_3 or SiO_2 at lower temperatures. However, under hot corrosion conditions, the properties of oxide scales are different. The protective oxide scale should not react with the corrosive environment and at the same time, it should not allow the corrosive species to diffuse into the coating. Therefore, development of a suitable coating is of great interest for such applications. Recent results have demonstrated the necessity of protective coatings with a higher degree of high temperature capability than the existing coating systems in order to prolong the life of superalloy components under hot corrosion conditions [26-30]. Coatings with sufficiently high aluminium contents to ensure protective and chemically inert alumina formation and re-healing after reaction with the corrosive environment can only satisfy these requirements. Literature review reveals that the platinum aluminide coatings perform well under high temperature oxidation and type I hot corrosion conditions, but the performance is poor under type II hot corrosion conditions, although the performance is superior to conventional aluminides. The choice therefore, appears to be MCrAlY based overlay coatings. The chromium rich and /or silicon containing MCrAlY coatings offer improved corrosion resistance at lower temperatures that are encountered in gas turbine engines. However, recent results have shown that Si containing MCrAlY coatings are detrimental to hot corrosion and life reduces significantly at higher temperatures. However, the MCrAlY coatings without Si offer good oxidation and hot corrosion resistance at higher temperatures. As mentioned earlier, the lives of the coatings are mainly limited by aluminium depletion occurring upon aluminium consumption as a result of alumina scale growth and repeated spallation / reaction and re-healing of the alumina scale during oxidation and type I hot corrosion conditions. Similar is the case with chromia scales under type II hot corrosion conditions.

Stringer has outlined the basic rules for proper selection of a coating for the given application [51]. The hot corrosion resistance of the aluminide coating has improved by the addition of yttrium. Swadzaba et al [52] have shown that regeneration of aluminide coatings on turbine blades, which have worked in highly corrosive environments, is possible to regain their initial properties. The recommended regeneration techniques are pack cementation for vanes and the slurry technique for rotating blades. The degradation of aluminide-coated first-stage blades in the engines of a light military helicopter operated by the Royal Australian Navy (RAN) was reported [53]. The different modes of degradation are foreign object damage, erosion, hot corrosion, coating cracking and coating alloy inter-diffusion. Small concentrations of NaCl, as low as 10 ppm in the gas stream with Na_2SO_4 cause hot corrosion more severe and attack is greater if NaCl is in solid state.

An overview on protective coatings for gas turbine applications, their role and design was given by Goward [54]. The performance of various pack, gas phase, and noble metal aluminide and overlay coatings applied on F100 first stage turbine blades under simulated service conditions were studied and estimated their relative lives [55]. Ni-Co-Cr-Al-Y overlay coating showed superior performance compared to other coatings in 300 hrs tests. The novel technique namely "Law of mixtures" for evaluation of mechanical properties of aluminide coatings over the temperature range 20-1000^0 C and the ability of the technique to predict the mechanical behaviour was reported [56]. The high temperature corrosion of a commercial aluminide coating on IN-738LC and MAR-M002 at 700 and 830^0 C was found to degrade with different mechanisms at two temperatures. Meethan reviewed various types of protective coatings used in the turbine and combustor sections of gas turbine engines operating in aero environments [57]. He described the factors affecting the coating selection for turbine rotor blade applications and the aluminide coatings continue to provide good and cost effective protection on many turbine blades. The several environmental effects on the mechanical properties of coated superalloys and comparative performance of various coatings are provided [58].

It is widely accepted that multi-layered and graded compositions can offer significant advantages over single layered coatings. A composite coating of aluminide-yttrium exhibits good stability and excellent corrosion resistance in a cyclic high temperature hot corrosion environment [59]. Yttrium addition helps in forming Y_2O_3 and (Y, Al) O-type compounds in grain boundaries of Al_2O_3 and NiAl effectively prohibits the fast diffusion of oxidants, thus helps to improve the hot corrosion resistance significantly. Improvement of hot corrosion resistance of the aluminide coating by the addition of yttrium has also been reported in another study. There is a work by a Japanese group showing that elimination of yttrium from MCrAlY has a positive effect on slowing down the TGO growth rate [60].

Platinum aluminide coatings are more resistant to hot corrosion than simple aluminides on land based gas turbines. The role of platinum in improving the hot corrosion resistance of pt-modified aluminide coating through electrochemical method revealed that platinum significantly improves the resistance to basic fluxing by increasing Al_2O_3 content in the platinum rich surface. Due to this, the difference in breakthrough potential between the two coatings was found to be 400 mV. The platinum free coating was found to corrode severely while coating with platinum showed slight corrosion after 310 hrs in free corrosion test. However, platinum modified aluminide coatings and simple aluminide coatings are having the same resistance to acidic fluxing.

Significant amount of work was carried out in developing the hot corrosion resistant MCrAlY bond coatings [8, 28, 61-62] and the lifetimes of various coatings including platinum aluminide, tested in number of engines is presented in Table VIII. The lifetime of MCrAlY system has been found to be improved considerably by aluminizing or incorporating platinum into the system. MCrAlY based bond coatings play a significant role in providing rough surface for the application of thermal barrier coatings and to provide protection for the alloy from oxidation and / or hot corrosion. Due to proven performance of these coatings for a variety of superalloys for over two decades in different applications, MCrAlY coatings have been studied extensively [28, 63-65]. It is important to mention that the durability of the MCrAlY coating also depends on the coating technique employed. The life of coating applied by arc ion plating is significantly lower as the attack occurs at the edges due to large variation in thickness i.e. about 6 times less than the coating thickness at the middle. Different degradation mechanisms proposed are oxidation followed by scale cracking and spallation, mixed oxidant attack, erosion and molten salt induced attack.

The effect of aluminium in forming a thermodynamically and chemically hot corrosion resistant layer and thereby improving the life of the components by reducing mass gain in MCrAlY coatings is very essential. Further, optimum content of aluminium in the coatings is extremely necessary to enhance their lifetime. In fact, it only decides the lifetime of coatings and hence the essentiality of aluminium reservoir in the coatings.

3.5. Optimization of aluminium content

Fig.12 illustrates the influence of aluminium on weight loss in the MCrAlY based coating model alloys [30]. It is very clear that aluminium plays a major role in affecting the hot corrosion resistance of MCrAlY alloys though the concentration of other alloying elements remains constant. The weight loss is maximum for the model alloy containing 6% aluminium, decreases with increase in aluminium content to 9% and minimum weight loss is observed for the alloy containing 12% aluminium. The behaviour is same for all the model alloys irrespective of cobalt content i.e. whether the cobalt is 10 % or 20% (Fig.12). Therefore, the minimum amount of aluminium required to be present in the MCrAlY based bond coatings is 9%. It is very important to mention that the optimum content of aluminium required for providing good hot corrosion resistance is 12%, though its effect is marginal when compared with 9% aluminium containing alloys [26-28].

It is also important to mention that the combination of cobalt and aluminium contents is a must to exhibit good hot corrosion resistance. The recent results clearly showed that cobalt plays a significant role in supporting aluminium to form a chemically and thermodynamically stable oxide scale. The optimum amounts of aluminium and cobalt are to be present in the MCrAlY bond coatings. Hence, the combination of cobalt and aluminium in association with chromium, yttrium and nickel provides good hot corrosion resistance [26-29].

Table 8. Lifetimes of coatings for aero, industrial and marine gas turbines [8]

System	Coating	Relative lifetime
Land based aircraft DS MAR-M200+Hf	Uncoated	<<0.5
	Standard aluminide (PWA 73)	1.0
	Rh-Al (BB)	<1.5
	Gas phase aluminide (PWA 275)	<1.5
	Modified aluminide (PWA 263)	1.5
	Pt-Al (RT 22)	1.5
	NiCoCrAlY overlay (PWA271)	2.75
Marine aircraft MAR-M002	Standard aluminide	1
	Pt-Al	2.5-3
	CoCrAlY	2
Industrial IN 738	Uncoated	1
	PtCrAl	>3
	CoCrAlY (plasma sprayed)	>2
	CoCrAlY (clad) plus Al diffusion coating	>6
	FeCrAlY	2
Marine Rene 80	Standard aluminide	1
	Rh-Al	1-2
	CoCrAlHf	2
	CoCrAlY	1
	CoCrAlY+Pt/Hf	2
	CoCrAlHfPt	2.3

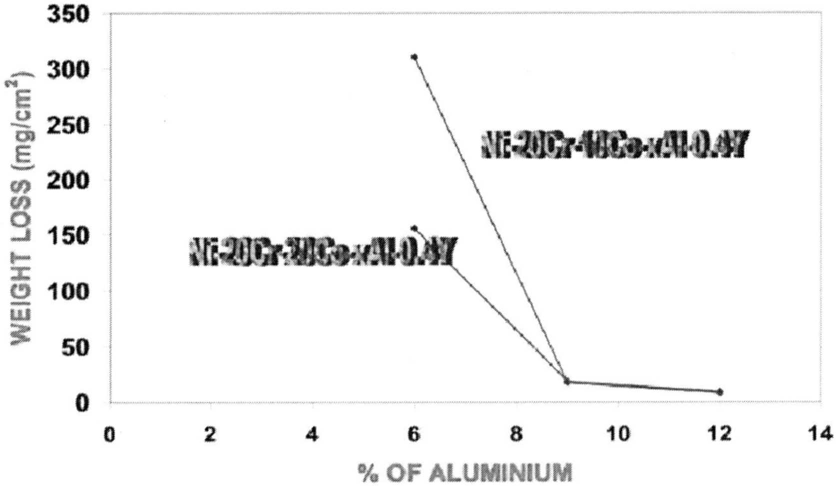

Figure 12. Effect of aluminium on hot corrosion of MCrAlY based coating model alloys [30]

3.6. Identification of new coating composition

The lives of different typical coatings applied on CM 247 LC alloy exposed at 900^0 C in NaCl and vanadium containing environments showed that the life of coatings varies with the composition and the environment (Table IX) [28]. The NiCoCrAlY coating exhibits maximum life both in sodium chloride as well as in vanadium containing environments. It is interesting to note the effect of absence of cobalt in NiCrAlY and nickel in CoCrAlY coating in reducing the lifetime as maximum lifetime is observed for the NiCoCrAlY coating in both the environments. It is clear that NiCrAlY and CoCrAlY coatings do not improve the lifetime, but the lifetime will be enhanced only if the coating contains Co in NiCrAlY and Ni in CoCrAlY coating. It is also observed that traces of silicon as well as hafnium reduce the coating life drastically (Table IX). It means that hafnium or silicon-containing compounds readily react with the corrosive salts at elevated temperatures and thereby reduce the coating life considerably [26-29]. Therefore, the ideal combination is 22% cobalt and 12% aluminium along with 18% chromium and 0.5% yttrium.

Table 9. Lifetimes of different MCrAlY type coatings in various evironments [28]

MCrAlY coating	Chloride environment (hrs)	Vanadium environment (hrs)
CM 247 LC alloy	<4	<2
Ni-16Cr-5Al-0.5Y	38	140
Co-30Cr-18Al-0.5Y	152	155
Ni-18Cr-22Co-12Al-0.5Y	>480	>300
Ni-18Cr-22Co-12Al-0.5Y + traces Hf	285	140
Ni-18Cr-22Co-12Al-0.5Y+ traces Si	35	158

To obtain maximum durability of a coating even with low-grade fuels with high pollutant content levels, an innovative coating should contain maximum aluminium reservoir together with high chromium content. Such coatings are usually MCrAlY type. A new MCrAlY coating was innovated by using electro-less technique [66-67]. The aluminium concentration was modified with the use of a modified aluminization and / or by introducing a diffusion barrier and the aluminium was changed by use of platinum aluminide coating. Using platinum aluminide altered chromium solubility [67].

In the new coating, a Ni-W diffusion barrier (5 μm thick), which is rich in W up to 50% mass, was introduced to prevent aluminium to diffuse from coating to the substrate. Over this layer 75 μm thick MCrAlY coating (NiCrAlYTa) was deposited by electro-less route. Subsequently 7 μm thick platinum and inward pack aluminizing was done (Fig.13). The hot corrosion tests with sodium sulphate showed improved performance over typical (SC/LPPS) MCrAlY / platinum aluminides and the innovative coating (SC / NiW + NiCrAlYTa + Pt + HA) maintained its integrity even after exposure for 1000 hours in pure sodium sulphate environment [67]. In order to prove the performance of a new coating, further experimentation with aggressive species like NaCl and V_2O_5 is extremely necessary.

Another new coating namely Ir-Ta applied by EB-PVD followed by pack aluminizing was developed for Ni-based superalloys. This coating exhibited good high temperature oxidation and hot corrosion resistance [68-69]. Ir has a high melting temperature (2716K), excellent chemical stability and low oxygen permeability, while Ta also has high melting temperature (3269K), chemical resistance to aggressive propellant gases and low diffusion rates in Ni-based superalloys. During the hot corrosion process, a high melting point compound $NaTaO_3$ (2083K) forms ($Ta_2O_5 + Na_2O = 2 NaTaO_3$), which is stable. Further, both α– Al_2O_3 and Ta_2O_5 (2163 K) form on the surface of coating and prevents inward diffusion of sulphur and oxygen and outward diffusion of chromium and nickel from the substrate, thereby eliminating the formation of sulphides which in turn enhances the hot corrosion resistance. The addition of Ir in platinum aluminide is also promising. Deposition of Pt-30at% and Pt-50at% Ir alloys followed by conventional aluminizing demonstrated enhanced oxidation and type I hot corrosion resistance [70-71]. Ir-Ta-Al coating appears to be another effective novel metallic bond coating materials for Ni-based single crystal superalloys. However, the conclusions are based on short-term tests in the presence of pure sodium sulphate. Hence, detailed research is needed in the presence of aggressive species like sodium chloride and vanadium pent oxide to prove its performance.

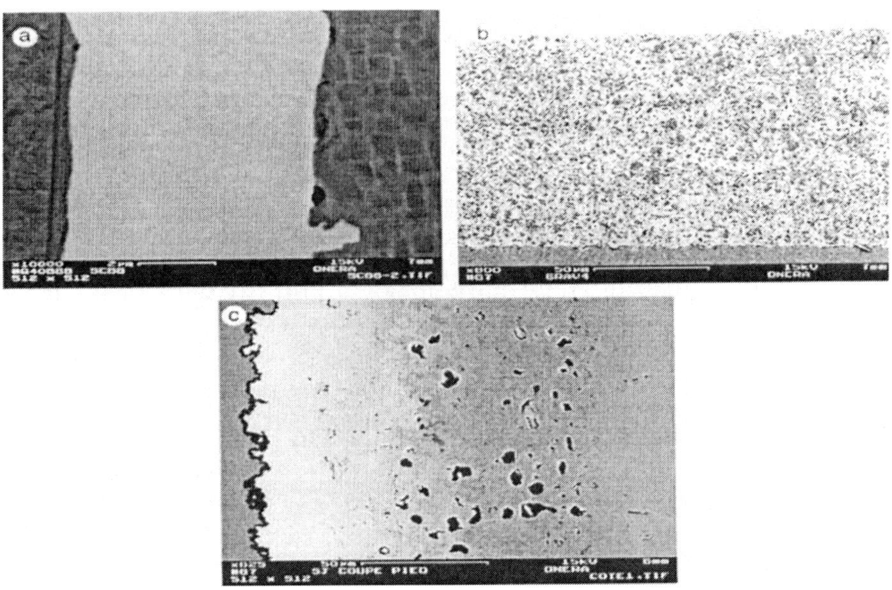

Figure 13. New coating deposit details a) 5 μm thick NiW layer b) 75 μm thick composite Ni-CrAlYTa c) 75 μm thick composite Ni-CrAlYTa + 7 μm thick platinum + inward pack aluminization [67]

Kawagishi et al have proposed an equivalent coating system for high temperature corrosion resistance of Ni- based superalloys [72]. This coating composition is designed to be thermodynamically in equilibrium with the substrate so that no inter-diffusion should occur between them. It is also reported that this coating technique helps in repair of turbine blades. Seraffon et al have attempeted to use combination of physical vapour deposition and magnetron sputtering technique to deposite Ni-Co-Sr-Al coatings of varied compositions and tested for their high temperature oxidation resistance [73]. However, hot corrosion tests have not been studied.

3.7. Influence of major and trace elements

As mentioned earlier, the NiCoCrAlY coating exhibits maximum lifetime among the coatings (Table IX) under type I hot corrosion conditions [28]. Neither NiCrAlY nor CoCrAlY coatings prolong the superalloy's life. In fact, CoCrAlY coating contains maximum amounts of chromium (30%) and aluminium (18%), but their positive effect is not observed. Similar is the case with NiCrAlY coating. It is also mentioned that traces of silicon and hafnium make the NiCoCrAlY coating highly susceptible to hot corrosion. Traces of silicon or hafnium modify oxide growth rate, the composition of oxide scale and consequently reduce the coating life significantly. The fluxing of oxide scale also becomes easy when traces of silicon or hafnium are present in the scale. The underlying mechanism is that hafnium and silicon are present in the grain boundaries of alumina scale leach out selectively by readily reacting with chlorine, vanadium, sodium and sulphur present in the environment to form corresponding compounds [28]. This results in dislodging the grains of alumina scale and creates instability of the oxide scale and thereby reducing the life of coatings significantly [28]. Further, the oxides of silicon and hafnium are soluble in molten basic sulphate and the basic fluxing dominates in the high temperature hot corrosion region (850-950^0 C). By observing the maximum life for the NiCoCrAlY coating which does not contain either silicon or hafnium at 900^0 C evidences it. The reaction mechanisms leading to reducing the life of coatings containing silicon or hafnium are given below:

$$HfAl_2O_3 + Na_2SO_4 + NaCl = NaAlO_2 + HfCl_4 + SO_3$$

$$HfAl_2O_3 + Na_2SO_4 + NaCl + V_2O_5 = NaAlVO_2 + HfCl_4 + SO_3$$

$$SiAl_2O_3 + Na_2SO_4 + NaCl = NaAlO_2 + Na_2SiO_3 + SO_3$$

$$SiAl_2O_3 + Na_2SO_4 + NaCl + V_2O_5 = NaAlVO_2 + Na_2SiO_3 + SO_3$$

The superior performance was confirmed by observing minimum corrosion affected zone for NiCoCrAlY coatings and more for other coatings [26]. It was further evidenced by observing a uniform, protective and adherent alumina scale on the surface of NiCoCrAlY coating (Fig.14). The above results clearly confirm that silicon addition is not beneficial for type I hot corrosion resistance rather it helps to combat type II hot corrosion [26]. Luthra et al and Bauer et al were reported that chromium [74] and silicon [75] respectively are particularly beneficial in coatings for protection against type II hot corrosion [76]. Reactive active element research related to oxidation characteristics and relevant mechanisms were the major works in overlay type coatings for the last one-decade [77]. However, the effect of reactive elements on hot corrosion resistance is still not well understood. Therefore, it may not be advisable to use silicon for protection of superalloys against both types of hot corrosion. Rather chromium addition is more appropriate to develop coatings for protection of components against hot corrosion (type I and II) as its presence does not have detrimental effect on type I hot corrosion.

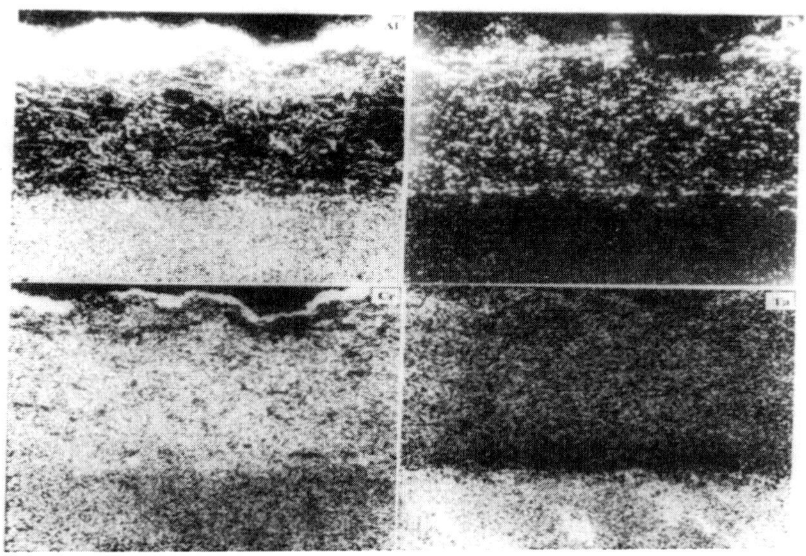

Figure 14. EPMA data showing the formation of continuous and protective alumina scale on coating surface [28]

3.8. Degradation mechanisms

Inter-diffusion of aluminides with the base alloys resulting decrease in life is the dominating degradation mechanism in diffusion aluminide coatings. Lower sulphur contents of diffusion coatings improves alumina scale adherence and the adherence of alumina scale is very good with platinum aluminide coatings. Though platinum improves the adherence and improves the life of coating, the degradation is same at the end. Clear understanding on the effects of platinum and various forms of active metals and their oxides will help to improve the use of aluminide coatings as well as bond coatings for thermal barrier coatings.

During service, coatings degrade at two fronts: the coating / gas path interface and the coating / substrate interface. Deterioration of the coating surface at the coating / gas path interface is a consequence of environmental degradation mechanisms. Solid-state diffusion at the coating / substrate interface occurs at high temperatures, causing compositional changes at this internal interface that can compromise substrate properties and deplete the coating of critical species. In the worst case, inter-diffusion leading to the precipitation of brittle phases can cause severe loss of fatigue resistance. The following section describes a novel mechanism namely an electrochemical phenomenon that explains MCrAlY based bond coating degradation process in detail.

If NiCoCrAlY coating is considered, hot corrosion takes place by oxidation of nickel, cobalt, chromium and aluminium at the anodic site (forms Ni^{2+}, Co^{3+}, Cr^{3+} and Al^{3+} ions) while at the cathodic site, SO_4^{2-} is reduced to SO_3^{2-} or S or S^{2-}. Since the metal ions i.e. Ni^{2+}, Co^{3+}, Cr^{3+} and Al^{3+} are unstable at the elevated temperature and therefore reacts with the sulphur ions to form metal sulphides. The metal sulphides can easily undergo oxidation at high temperatures and form metal oxides by releasing free sulphur ($MS + 1/2\ O_2 = MO + S$). As a result, sulphur concentration increases at the surface of the coating and enhances sulphur diffusion into the coating and forms sulphides inside the coating. The practical observation of sulphides in hot corroded components clearly indicates that electrochemical reactions take place during the hot corrosion process [26, 28]. Similar reactions take place in vanadium environments except the formation of oxides at cathodic sites. Fig.15 illustrates an electrochemical model showing the NiCoCrAlY degradation is electrochemical in nature.

Therefore, the hot corrosion of NiCoCrAlY coating can be considered electrochemical in nature and the relevant electrochemical reactions are shown below:

At the anode	At the cathode
$Ni = Ni^{2+} + 2e^-$	$1/2\ O_2 + 2e^- = O^{2-}$
$Cr = Cr^{3+} + 3\ e^-$	$SO_4^{2-} + 2e^- = SO_3^{2-} + O^{2-}$
$Co = Co^{3+} + 3e^-$	$SO_4^{2-} + 6e^- = S + 4O^{2-}$
$Al = Al^{3+} + 3e^-$	$SO_4^{2-} + 8e^- = S^{2-} + 4O^{2-}$

Similar mechanism is applicable to MCrAlY coatings family.

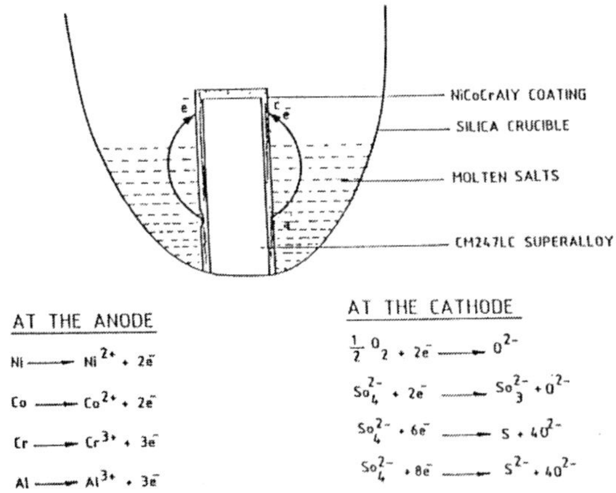

Figure 15. An electrochemical model showing degradation of MCrAlY type coatings is an electrochemical phenomenon [26]

The motivation behind suggesting an electrochemical model is to show that degradation of materials (alloys or coatings) in molten salts is electrochemical in nature and thereby to develop a rapid hot corrosion test, which will be extremely useful to select the most potential candidate materials. The normal weight gain method requires hundreds of hours to test one material whereas the electrochemical method needs hardly 4 to 5 hours. It is a major advantage of electrochemical techniques over conventional weight change methods. Electrochemical monitoring of corrosion by linear polarization, coulostatic and A.C. impedance techniques have been established and widely used in aqueous solutions. On the contrary, the evaluation of corrosion resistances of various metals and alloys in molten salts are usually conducted by weight loss methods. Since the degradation of superalloys [6] and MCrAlY coatings [26,28] is shown as an electrochemical process, same as in aqueous solution, the electrochemical techniques should be quite helpful not only in evaluating coatings for hot corrosion resistance, understanding their hot corrosion mechanisms but also for developing hot corrosion resistant coatings for the superalloys [6,28].

3.9. Ceramic coatings

Thermal barrier coatings (TBCs) are finding increased application on overall component design. Over the past 25 years, cooling technology has contributed roughly 370^0 C (700 F) (from solid blades to advanced film cooling) in turbine temperature capability; further advances may be achieved with even more sophisticated cooling schemes. Superalloy material and processing advances (from equiaxed crystalline structure to third generation single crystal) have added approximately 120^0 C (250 F). However, superalloys now operate in some applications at 90% of their melting point. TBCs have the potential to reduce the substrate temperatures further with improved current production methods.

Application of zirconia thermal barrier coatings over diffusion aluminide and / or MCrAlY bond coatings is known to increase the life of components further. The function of the ceramic topcoat is to insulate the metallic substrate from high surface temperature, thereby lower the components temperature and reduce the oxidation and hot corrosion of bond coatings. The ceramic coating can lower the temperature of a cooled blade by up to 170^0 C (306^0 F), while simultaneously reducing cyclic thermal strains and has gained prominence in coating technology. ZrO_2 stabilized or partially stabilized by the addition of MgO, CaO, Y_2O_3 is found to exhibit good performance because of its high coefficient of thermal expansion and very low thermal conductivity. It has been demonstrated that columnar structures applied by EB-PVD (Fig.16) has a significant improvement in life over plasma sprayed coatings [46]. The early work used EB-PVD ceramic coatings on NiCoCrAlY bond coating applied by low-pressure plasma. The latest trend is to use platinum aluminide as bond coatings. EB-PVD is often favoured over plasma deposition for TBCs on turbine airfoils since it provides a smooth surface of better aerodynamic quality with less interference to cooling holes [47]. However, the widely used plasma spray process has benefits, including a lower application cost, an ability to coat a greater diversity of components with a wider composition range and a large installed equipment base. The established ceramic topcoat for effective prevention of high temperature corrosion is the use of ZrO_2 partially stabilized with 8 wt% Y_2 O_3. This coating exhibits satisfactory performance for aero gas turbines and primary degradation mechanism is due to bond coating oxidation. An encouraging trend was observed by chemical vapour deposition of Al_2O_3 over NiCoCrAlY bond coating surface prior to overlaying a ZrO_2-$8Y_2O_3$ topcoat to the cyclic life of partially stabilized zirconia by suppressing the bond coating oxidation effectively.

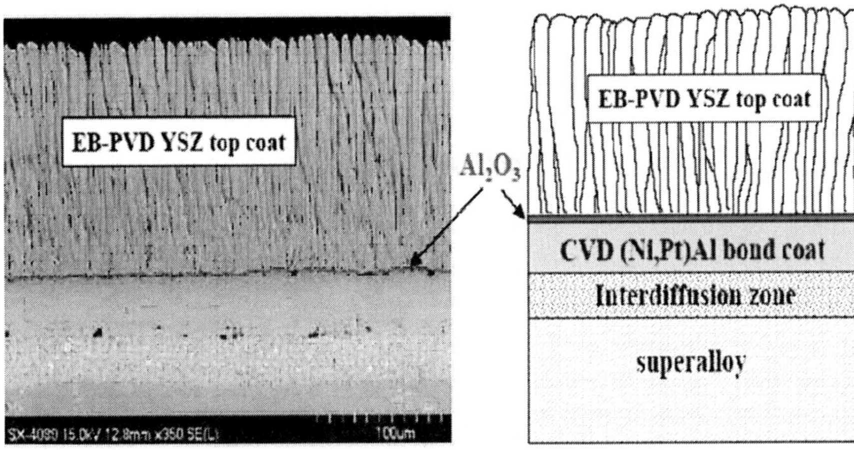

Figure 16. TBC structure deposited by EB-PVD over (Ni/Pt) Al bond coating [47]

However, it is not useful to marine gas turbines or industrial gas turbines because yttria (Y_2O_3), which is the stabilizer for zirconia reacts with vanadium present in many industrial quality fuels or $NaVO_3$ and forms YVO_4 [31]. Vanadium depletes the Y_2O_3 stabilizer from ZrO_2 matrix and thus destabilization of zirconia (i.e. transformation of zirconia from tetragonal and/or cubic to monoclinic phase upon cooling, which is accompanied by a large volume change) takes place. Destabilization of zirconia eventually causes delamination and spalling of the ceramic coating. In addition, molten salt can penetrate into the YSZ through pores and cracks (present in YSZ) and react with the metallic bond coatings. As a result, various corrosion products are generated, leading to crack formation and propagation within the oxide scale. In order to combat this problem, the pores and cracks should be sealed either by using lasers or application of an additional protective layer. Silicates like mullite ($3Al_2O_3.2SiO_2$) or cordienite ($2MgO.2Al_2O_3.5SiO_2$), CAS ($CaO.Al_2O_3.5SiO_2$), BAS ($BaO.Al_2O_3.2SiO_2$) and calcium silicate (1.8 $CaO.SiO_2$) applied over ceramic TBC helps in improving the hot corrosion resistance due to formation of high melting point sodium-alumino-silicates (>1573K) [78-85]. The rate of destabilisation of zirconia was restrained by the presence of additional over layers. However, as the pores can be present in over layered coatings, the molten salt can easily penetrate and interact with zirconia to form YVO_4 but the time required to reach the ceramic layer is higher. This aspect led to identifying CeO_2, an effective alternative to Y_2O_3 for marine and industrial gas turbine applications. Alternative stabilizers to ziconia are needed to further prolong the life of TBC coatings.

Knowledge that active elements in Ni-based superalloys and / or platinum modifications caused improved scale adherence in diffusion aluminide coatings might have led to the use of these coatings as bond coatings. The large beneficial effects possible in gas turbine efficiency by the use TBCs have sparked a resurgence in research on all aspects of scale adherence and failure mechanisms involved in loss of ceramic layers or reactions with hot corrosive salts. It was expected that future research will yield incremental improvements in scale adherence towards the goal of perfection required to attain prime reliant status necessary to take full advantage of saved cooling air. Research in this direction led to the application of TBCs directly on an advanced yttrium containing superalloy without the need for a bond coating [86-87].

In practice, the inlet gas temperature in the state-of-the-art gas turbines has already reached 1773 K, which exceeds the melting points of Ni-based superalloys used for gas turbine blades and vanes. Therefore, the necessity of improving the performance of TBC is highly essential. In addition, the coating degradation takes places due to erosion as well. Therefore, it is essential to apply multilayered coatings in a perfectly ordered manner to protect the components from all attacks / damages. It is worthwhile to concentrate on the development of appropriate coating system (multilayered) including thickness for different types of gas turbine engines. Optimization of coating process taking all the experimental parameters into account is another area of critical importance.

3.10. Life prediction modelling

Because of the dependence of blade integrity on TBCs, considerable efforts are necessary to develop life prediction models to aid the gas turbine design. It is important to validate the developed model with experimental data obtained both in the laboratory as well as in the field. Development of a suitable laboratory test for predicting the life of developed materials / coatings accurately in service is an important task for the Corrosion Engineers. Recently, Renusch and Schutze [88-89] have developed a prototype software to predict the lifetimes of yttria stabilized zirconia (YSZ) top coating as a function of a bond coating temperature for a given input of hot dwell time and temperature gradient and the results have been validated with laboratory data. Attempts were made to find a solution for the transient temperature transfer problem in bare and thermal barrier coated alloys with the computational fluid dynamics (CFD) part of analysis by application of ANSYS Fluent code receiving the temperature field of combustion gas, whereas computational structural mechanics (CMS) part concerning the temperature distribution inside the turbine blade was done by ABAQUS [90-94]. Degradation modelling is most important to coating design and development. Process modelling is most important to coating manufacture. Coating life and inherent substrate environmental resistance are key determinants in setting the intervals for engine inspection and overhaul. Only few models exist in the public domain that addresses any of these needs.

It is pertinent to note that predicting the lifetimes of same YSZ top coating under hot corrosion conditions has not been attempted yet. In fact, it is a complex and challenging aspect as software development involves taking into account various reaction mechanisms that takes place under a variety of service conditions [95]. Serious attempts are needed to develop such software, which will have tremendous improvement in enhancing the efficiency of various advanced gas turbine engines. As mentioned above, the results have to be validated with both laboratory and field data and achieve an excellent agreement.

3.11. Reliability

Reliability is a critical design factor that is needed for further development if TBCs are to be fully exploited to increase turbine efficiency [96-97]. TBCs fail as a result of erosion, impact damage, and interfacial oxidation of the bond coating or thermo-mechanical strain at the ceramic / metal interface. These factors and process variability, more than any other design factor, has slowed the introduction of these coatings for turbines. Improved understanding of interfacial behaviour is required to control coating properties and to predict performance. A more compatible and high temperature corrosion resistant bond between the TBC and either the metallic substrate or the bond coating requires continued long-term emphasis [98]. The processes by which TBCs are currently applied, namely air plasma spray and EB-PVD will likely continue to be the major manufacturing methods.

Engineers currently rely heavily on visual inspection to assess the condition of coated structures [97]. As a result, in-service condition monitoring and repair decisions focus on deterioration at the coating / gas path interface. Improved non-destructive evaluation methods would provide information on when the coating has to be removed and on the extent of base metal attack. An important need for repair of industrial gas turbine components is industry wide repair specifications and regulations of the quality of repairs.

The most effective method to achieve a consensus of all interested parties remains unclear. Methods to make local repairs of coatings are needed both during manufacture and operation. Also required are standard, industry-accepted methods to determine the durability and properties of refurbished coatings.

Aircraft engine repair needs are parallel to those for the industrial turbines, with the further complexity that these coating structure systems include more advanced designs and materials that tend to limit repair options. Incorporation of better models and data from condition-monitoring sensors will improve repair / replacement decisions. Still unclear is the extent to which many of the advanced coatings, such as TBCs, lend themselves to repair. Although some TBC overhaul is currently done, the extent to which TBC coated components can be repaired and re-used has never been fully determined.

Exploiting existing and advanced non-destructive evaluation (NDE) methods can aid significantly in developing and qualifying coating systems, improving process control during coating operations and characterising the integrity of coated structure during turbine engine manufacture, in-service condition monitoring and repair and overhaul operations [98-101]. Since each of these applications has specialised requirements, no single NDE technique will likely to serve all purposes. Development of programs for advanced NDE methods should focus on supporting these key areas with the goal of bringing the new methods into practice. The highest priority for further NDE development should be for those non-contacting methods that can examine the interior structure of the coating system, such as the coating / substrate interface and base metal. For aircraft engines in particular, the development of advanced NDE techniques and cost effective models will be essential for the assessment of components that are expected to be multi-wall or thin-wall structures with multilayered coatings used as an integral part of the component design and manufacture.

Future generations of higher-performance aerospace turbine structures will require newer materials, because in service superalloys are approaching the upper limit of their inherent temperature capability as mentioned earlier. Potential candidate materials to replace superalloys are inter-metallic compounds, monolithic and composite ceramics and refractory alloys. In addition, advanced cooling concepts will result in processing modifications and more complex cooling paths to meet the demands of advanced component designs. Most advanced materials and design modifications may result in component structures with inherently less resistance to aggressive environmental attack than current superalloys.

An important element that must be considered in innovative coating systems is the stability of structural components in the high temperature environment of an operating engine. The stability includes the long-term ability of the material to maintain its initial mechanical properties, chemical and micro-structural morphology. Research is needed on continuously graded coatings. Graded coatings like functionally graded materials (FGM) and nanostructures, offer potential advances in coating performance. The need also exists for ceramic coatings that can withstand higher temperatures. Graded coatings may demand alternative materials as well as alternative applications processes. Most significant in this area would be critical assessment of the use of FGMs as coatings and definition of the influence of multilayer and nanostructure morphology on resulting properties. However, the inherent in-homogeneity in the microscopic scale of these materials raises questions of high temperature stability that must be answered to establish the viability of these approaches. Advanced substrates like composites, ceramics and inter-metallic compounds will possess coating ability characteristics (e.g., diffusion rates and surface chemistry) very different from current superalloys. Novel concepts may be needed for coatings and surface treatments to protect these substrates. In essence, advanced processes and intelligent coatings are needed to manufacture future gas turbine engines for exhibiting ever greater efficiency.

4. CORRELATION BETWEEN LABORATORY AND SERVICE EXPERIENCE

Performance evaluation of materials / coatings in actual gas turbine engines is hardly practicable. Therefore, several simple methods were developed to asses the hot corrosion resistance of possible materials for blades and vanes. A satisfactory test should yield a prediction of service performance, with an estimation of component lifetime as a desirable objective. Attempts were made earlier to compare the results of laboratory corrosion tests with service experience. It is important to note that the corrosion of gas turbine components is greatly influenced by fuel and air quality. Generally, highly rated engines use relatively pure fuels and the air purity is also good unless move in marine environments.

In general, a more complete gas turbine engine environment can be simulated in burner rigs and the ranking of materials / coatings are in good agreement with the engine experience. The crucible test results are in agreement with burner rig test results in ranking the alloys. On the basis of corrosion products identification, it is concluded that burner rig test give the most realistic reproduction of corrosion mechanisms found in service, although furnace test as well as crucible tests could give satisfactory results. The burner rig and electrochemical tests produce similar corrosion morphology and corrosion rates are in good agreement with each other [22]. The performance of materials tested in electrochemical tests behaved in a similar manner in actual engines.

General Electric developed an advanced burner rig test cycle for studying the combined effects of oxidation and hot corrosion on gas turbine materials [22]. This test combines the cyclic elements of an aircraft engine operation cycle with the fundamental elements of the standard oxidation and hot corrosion tests. A dilute sea salt solution is atomized inducted into one atmosphere burner rig to simulate hot corrosion conditions present at coastal airports. Specimens are romped to oxidizing conditions similar to those present during take off and climb. The thermal gradients present along the length of the test specimens simulate those present on actual aircraft engine hardware and produce areas of oxidation, hot corrosion and combined hot corrosion / oxidation attack. This burner rig test was reported to be successful in reproducing the accelerated combined hot corrosion / oxidation attack observed in factory engine testing. The ranking of substrate / coating obtained with this method vary greatly from those obtained with conventional test methods. This is due to the fact that the ranking recorded in advanced burner rig test is a function of location on the pin and distinctly different rankings can occur in each attacked region.

The advanced burner rig test was used to test tube specimens that can be internally cooled during exposure to the combustion system. Several thermal gradients can be established across specimen walls like actual turbine hard ware by using this method. With the advanced burner rig test, it is possible to compare life times of TBC coatings with the function of thickness [22]. Nevertheless, significant surprises in terms of actual versus predicted coating performance continue to occur in the field. As a result, new laboratory tests or improved existing procedures are needed in real world applications.

5. SMART COATINGS

5.1. Concepts

A primary requirement for a coating to use in high temperature applications is slow growth rates and optimum adherence of the protective scales that forming on the metallic coatings during high temperature exposure. As the gas turbine engines experience both types of hot corrosion as well as oxidation and hence the need to develop smart coatings to effectively combat high temperature corrosion problems for obtaining maximum efficiency.

If a single coating can operate successfully over a range of temperatures with different forms of corrosion attack like type I and II hot corrosion and high temperature oxidation, the coating essentially respond to local temperature in such a way that it will form either alumina or chromia protective scale as appropriate. High purity alumina scales offer best protection against high temperature oxidation and type I hot corrosion and chromia scales against type II hot corrosion. Ideally a single coating should satisfy both the requirements. It is possible for a coating only if it contains chromium and aluminium rich graded coatings (Fig.17). The base coating should be a standard MCrAlY coating enriched with aluminum at its outer surface and a chromium rich layer at its inner surface. Under high temperature oxidation and type I hot corrosion conditions, the outer layer of the coating forms alumina scale, which provides protection and it, offers less protection under low temperature conditions. Under type II hot corrosion conditions, chromium rich layer forms chromia scale at a faster rate and provides protection. Thus, the smart coating can provide optimum protection by responding suitably to the temperatures that are encountered under actual service conditions of gas turbine engines. This optimum protection is possible because of the formation of the most suitable protective oxide scale in each temperature range of operation envisaged. In this sense, the coating responds to its environment in a pseudo-intelligent manner and hence the name **SMART COATING** [102-103]. The smart coating permits operation of gas turbine engines over a wide range of temperatures successfully for more than the designed life and helps in enhancing the efficiency significantly by effectively preventing oxidation, type I and type II hot corrosion. Such coatings were developed by the principle author and recommended to use for marine, industrial and aero gas turbines (Fig.18) [102, 104-105].

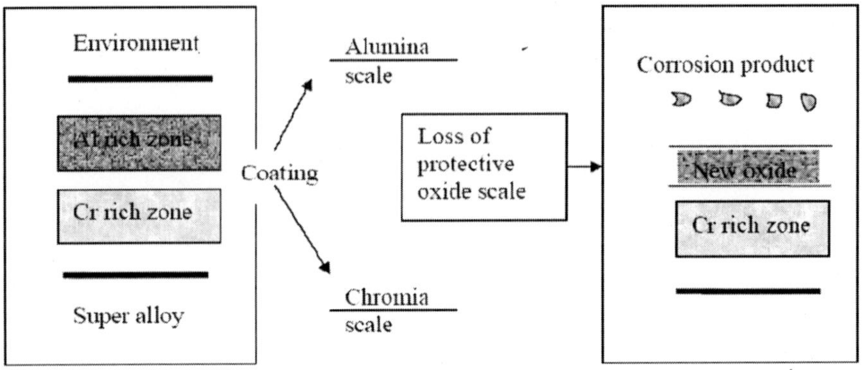

Figure 17. Schematic representation of response of a smart coating [102, 103]

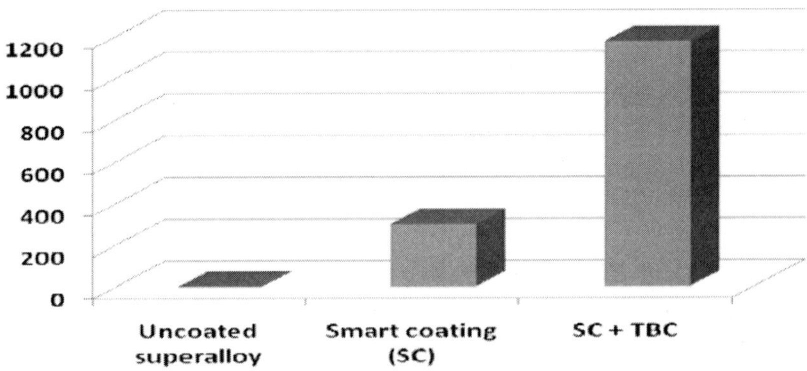

Figure 18. The performance of a Smart Coating (SC) [102].

5.2. Preparation

Preparation of smart coatings is really a challenging task. Therefore, selection of suitable surface engineering techniques to produce a quality coating is extremely essential. Here, selection of a single technique may not be of any help, but a combination of techniques is quite useful [102]. Further, the order of usage of selected techniques is also important as the microstructures of coatings vary and play a significant role in enhancing the life of components. The selection of technique for coating preparation should be based on the parameters that are controllable to get required microstructure. As mentioned earlier, the smart coating is a graded coating consisting of different zones (Fig.19) [102]. Each zone has its specific microstructure with definite composition and extends the life of gas turbine engine components significantly. Extensive research is needed both at the laboratory and field in order to establish the appropriate coating compositions, thickness, microstructures, coating techniques and superior performance of new smart coatings over conventional coatings.

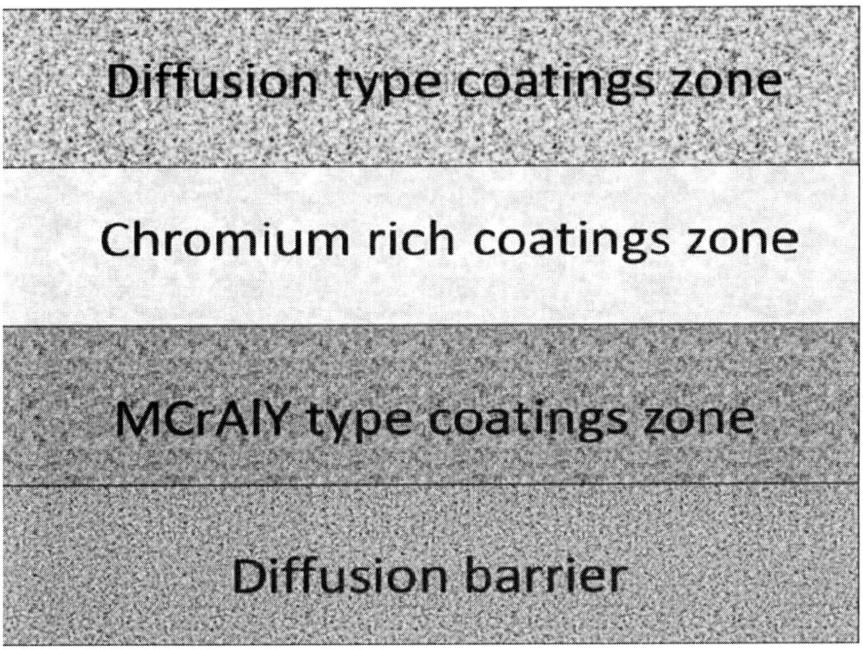

Figure 19. A schematic representation of microstructure of a smart coating [102]

6. MECHANICAL PROPERTIES OF COATED SUPERALLOYS

The mechanical properties of superalloys (especially cast alloys, which have their optimum mechanical properties in the as-cast condition) may deteriorate during the coating process (e.g. on thermal cycling during coating, during post-coating heat treatment, or by coating surface effects). The degradation of mechanical properties of the base material may lead to an increase in creep rate, thereby the elongation that can be tolerated before the coating starts to crack will be reached in a shorter time. In a corrosive environment, stresses are enhanced due to corrosion of coating and result in a measurable reduction of the cross section or by selective corrosion attack after a crack is initiated in the coating and propagates rapidly along grain boundaries or by both general and selective attack.

With the solid blades that used to be installed in turbines, the interaction between coating and base material was not significant because the potential influence on the load-bearing cross-section was very less. However, today's channelled, cooled blades have thin walls and a substrate-coating interaction over a depth of only few tenths of a millimetre can have a significant effect. Such interactions have been investigated in detail and reported [106]. It is important to mention that the coating-substrate interaction complicates the task of determining the various factors that are involved. The effectiveness of a protective coating for a particular material under service conditions can therefore be assessed only by techniques in which all the parameters like substrate, coating and environment as shown in Fig.20 are taken into account [40].

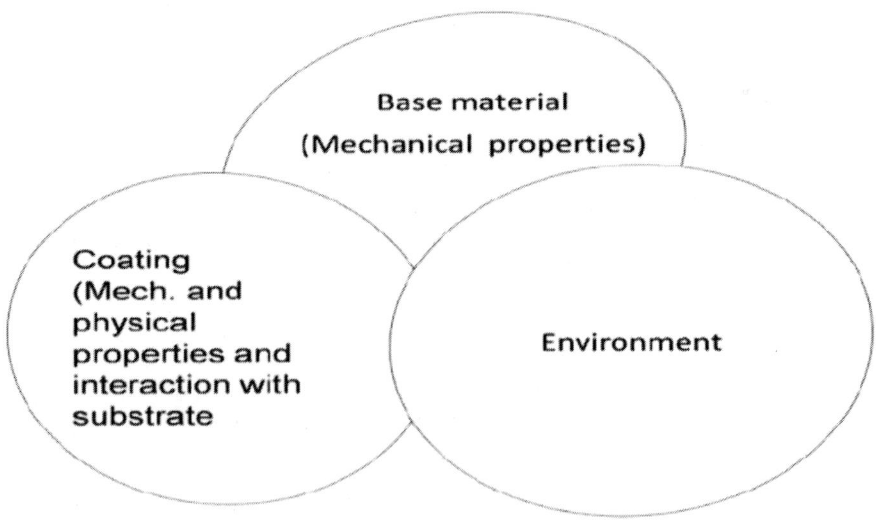

Figure 20. Factors affecting the coated superalloy components [108]

Swindells [107] and Cooper and Strang [108] have shown that one of the major problems faced by all coating systems is their mechanical properties. To achieve a good coating without discontinuity at the interface some inter-diffusion is required, but, if this diffusion occurs to any significant level during operation, the coating will slowly be degraded. The work on diffusion mechanisms is important in determining the long-term stability of the coatings. However, it has particular significance when discussing the mechanical properties of coated components, reviewed by Strang [109] and Czech et al [110].

The major problem is how to compare the properties of a coated component with an uncoated component. When such comparisons are made, it is of great importance how the cross section is measured, whether comparisons are made on the dimensions of the section before coating or whether the total thickness of base metal and coating are compared with a similar thickness of uncoated metal. This choice dominates the results in thin-section components, particularly when the coating has very different mechanical properties from the underlying metal. It effects the results obtained in creep, fatigue and thermal fatigue tests. However, the problem of the effect of coating properties on the mechanical integrity of the component is only of interest in some cases. For instance, as pointed by Goward [111] if failure is by thermal fatigue as in aircraft turbines, then the surface coating effects on mechanical stability are important; if failure is by hot corrosion, as in industrial and marine turbines, then the effect on mechanical properties is generally of a second order compared to the protection afforded by the coating.

7. SUMMARY AND FUTURE RESEARCH TRENDS

7.1. Summary

High temperature corrosion i.e. oxidation, type I and II hot corrosion are the degradation modes for the gas turbines. Hot corrosion is a major problem for the gas turbines engines due to which failures take place during service. Though advanced superalloys comprising new alloying elements such as rhenium, ruthenium and iridium (4th and 5th generation) that have been developed recently, exhibit considerably improved high temperature strength properties, their hot corrosion resistance is found be very poor. Therefore, there is a need to apply appropriate hot corrosion resistant coatings on the superalloys as the gas turbine engine components should exhibit both high temperature strength as well as hot corrosion resistance. Considerable amount of research has been carried out in the coatings area. As a result, new compositions, graded coatings have been emerged and efforts have also been made to predict their lives. However, no work was reported to predict the lives of coatings under hot corrosion conditions. Recently, smart coatings with varied techniques and compositions have been reported, which provides total protection to the superalloy components against high temperature oxidation, type I and II hot corrosion with their intelligent behaviour which in turn enhances the efficiency of gas turbines by eliminating failures during service. Therefore, it is recommended to apply smart coatings for the advanced superalloys as bond coatings used in all types of modern gas turbines i.e. aero, industrial, marine in order to obtain ever greater efficiency, which is essential in the present world.

7.2. Future research trends

No established ISO standard procedure is available till now for evaluation of materials and coatings for their hot corrosion resistance and hence the need to establish such standards in near future. In fact, the immediate concentration should be focused on this issue. Development of new alloys with a combination of experimentation, modelling and black art is necessary as further improvements in Ni-based superalloys are less possibility because the superalloys now operate at about 90% of their melting temperature. The future generation of high performance aerospace structures or potential candidate materials to replace superalloys are intermetallic compounds, monolithic and composite ceramics, refractory alloys and research is needed in this direction. The developmental work should take place concurrently with coating system. Improved understanding of interfacial behaviour of TBCs and a more compatible hot corrosion / oxidation resistant bond coatings either by modification of chemistry or more stringent control of undesirable elements to control coating properties and to predict their performance is highly essential. Hence, the development of constructive life prediction modelling under simulated gas turbine engine conditions is necessary. Improved on-line control to ensure reproducible coated structures and within the service limits is also needed as increased improvements to current coating technologies are unlikely to meet the goals of future generation high performance turbine engines. Then, development of Smart Coatings to combat type I & type II hot corrosion and high temperature oxidation for gas turbines is a challenging task to the Corrosion Engineer and developmental work in this area has lead to some smart coatings. Further research is highly essential for their understanding and to develop alternative smart coatings. Extensive research is needed both at the laboratory level and field to optimize coating composition, thickness, microstructure, identification of appropriate surface engineering techniques, and their priority of use and to prove their performance for manufacture of advanced gas turbine engines for exhibiting ever-greater efficiency.

8. ACKNOWLEDGEMENTS

The principle author (IG) is grateful to Japan Society for Promotion of Science (JSPS) for providing financial assistance for his stay at National Institute for Materials Science, Tokyo, Japan.

REFERENCES

1. www.salvatoreaiello.com/main.shtml
2. M. Konter and M. Thumann, J. Mater. Process Technol., 2001, 117, 386-390
3. M.R. Khajavi and M.H. Shariat, Engineering Failure Analysis, 2004, 11, 589-597
4. J.M. Gallardo, J.A.Rodriguez and E.J. Herrera, Wear, 2002, 252, 264-268
5. N. Eliaz, G. Shemesh and R.M. Latanision, Engineering Failure Analysis, 2002, 9, 31-43M.
6. I.Gurrappa, Oxid. Met., 1999, 50, 353-382
7. C.J. Wang and J.H. Lin, Mater. Chem. Phys., 2002, 76, 123-129
8. S.R.J.Sandrs, Mat.Sci.Tech, 1986, 2, 282-285.
9. P.Sterimtz,C. Duret and R.Morbioli, Mate.Sci.Tech, 1986, 2, 262-271.
10. R.A.Papp, J.H. Devan,D.L. Druglass,p.C. Nordine, F.P. Pettit and D.P. Whiltler, Matr.Sci.Engg, 1981, 50, 1-17
11. J.F.G.Conde, E.Erdos and P.Huber, High temperature Alloys for Gas Turbines (eds.) R.Brunetaud et al., D.Reidel publishing Co., The Netherlands (1982) p 99
12. F.S. Pettit and G.H.Meurer, Super alloys 1984, 651-687
13. R.A.Rapp, Pure & Appl.Chem., 1990, 62, 113-122

14. N.Otsuka and R.A.Rapp, J.Electrochem. Soc., 1990, 137, 46-52

15. V.S.Bhide and W.W. Smeltzer, J.Electrochem. Soc., 1981, 128, 902-908

16. F.C.Yang and D.P. Whittle, Corr. Sci., 1983, 23, 285-290

17. K.Natesan, Corrosion, 1985, 41, 646-655

18. K.Natesan, Materials for Coal Gasification (ed.,) W.T. Bakker, S. Dapkunas and V.Hill, American Soc. for Metals, Metals Park, Ohio, (1988) 137-146

19. M.P.Srivastava, T.H.Rao and R. Sivakumar, Trans.IIM., 1985, 38, 357-361

20. P.Hancock, Corr. Sci., 1982, 22, 51-65

21. A.Anderson,B.Haflan, P. Kofstad and P.K.Lillerud, Mater. Sci. & Tech., 1987, 87, 45-50

22. J.A. Cinner and W.B. Connor, J. Met., 1994, 21, 35-41

23. E.L.Simons, G.V.Browing and H.A.Liebhafskers, Corrosion, 1955, 11, 505-511t

24. I.Gurrappa, J.Mater.Sci.Eng., A 2003, 356, 372-380

25. I.Gurrappa, Oxid. Met., 2003, 59, 321-340

26. I.Gurrappa, Mater.Sci.Technol., 2003, 19, 178-185

27. I.Gurrappa, J.Mater.Sci.Lett., 2001, 20, 2225-2229

28. I.Gurrappa, Surf.Coat.Technol., 2001, 139, 272-280

29. I.Gurrappa, Mater. Manuf.Process, 2000, 15, 761-769

30. I.Gurrappa, J.Mater.Sci.Lett., 1999, 18, 1713-1721

31. I.Gurrappa, J.Mater. Sci.Lett., 1998, 17, 1267-1272

32. I.Gurrappa, J.High Temp.Mater.Sci., 1997, 38, 137-144

33. I.Gurrappa, Corr.Prev.Control., 1997, 44, 151-178

34. I.Gurrappa, J.Electrochem.Soc., 1999, 48, 187-195

35. I.Milosev and B. Navinsek, Surf. Coating & Tech., 1994, 63, 173-180

36. X.Zhang and R.A.Rapp, J.Electrochem.Soc., 1993, 140, 2857-2863

37. Y.S.Hwang and R.A. Rapp, J. Electrochem. Soc., 1990, 137, 1276-1281

38. T.Kobayashi, Y.Koizumi, S. Nakazawa, T. Yamogata and H. Harada, Proceedings of Fourth International Charles Parsons Turbine Conference on "Advances in Turbine Materials, Design and Manufacturing", November 1997, New Castle, UK pp 766-773

39. P.Canon and T. Khan, Aerospace Sci. Technol., 1999, 3, 513-521

40. I.Gurrappa, I.V.S. Yashwanth and A.K.Gogia, The selection of materials for marine gas turbines, Gas Rurbines ISBN:979-953-307-816-7, Volkov Konstarton (Editor), 2012, INTECH Publishers, USA, p 51-70

41. I.V.S. Yashwanth, I.Gurrappa and H.Murakami, J. Surf. Engg. Mater. Adv. Technol., 2011, 1, 130-135

42. K. Matsugi, Y.Murata, M.Morinaga and N. Yukawa, Proceedings of 7th International Symposium on Superalloys ed. By S.D.Antolovich et al. TMS, Warrendale, PA, (1992) p 307

43. Md.Moniruzzaman, Y. Murata, M.Morinaga, N.Aoki and T. Hayashida, ISIJ International, 2002, 42, 1018-1125

44. Y.Murata, S.Miyazaki, M.Morinaga and R.Hashizume, "Hot corrosion resistant and high strength Ni-based single and directionally solidified super alloys developed by the d-electrons concept", Super alloys 1996 ed. by R.D.Kissinger, D.J.Deye, D.L.Anton, A.D.Cetel, M.V.Nathal, T.M.Pollock and D.A.Wood ford, The Minerals Metals & Materials Society 1996

45. I. Gurrappa, A. Wilson, H.L.Du, J.S. Burnell-Gray and P.K. Datta, ECS Trans., 2008, 6, 67-75

46. H.W. Grunling, H. Rechtanbacher and L. Singheiser, Mater. Sci. Forum, 1997, 251-254, 483-502

47. Hancock, P., Malik, M., "Coating Systems and Technologies for Gas Turbine Applications", Materials for Advance Power Engineering, Part 1, 1994, pp. 685-704

48. S.Kuroda, J.Kawakita, M. Watanabe and H. Katamoda, Sci. Tech. Adv. Mater., 2008, 9, 33002

49. G.W.Goward and L.Cannon, J,Eng.Gas Turbines Power, 1998, 110, 150-154

50. G.W.Goward and L.L.Seigle, ASM Handbook, Surface Engineering, 1994, 5, 611-620

51. J.Stringer, Mater.Sci.Eng., 1987, 87, 1-10

52. L.Swadzaba, B.Formanek and A.Maciejny, Meter.Sci.Eng.,A 1989, 121, 407-412

53. J.A.Cocking, N.A.Burley and G.R.Johnston, High Temp.Tech., 1986, 4, 175-180

54. G.W.Goward, Surf.Coat.Technol., 1998, 108-109, 73-79

55. A.J.A.Mom and H.J.C.Hersbach, Mater.Sci.Eng., 1987, 87, 361-367

56. P.Hancock, H.H.Chien, J.R.Nicholls and D.J.Stephenson, Surf.Coat.Tech., 1990, 43/44, 359-370

57. G.W.Meethan, Mater.Sci.Tech., 1986, 2, 290-294

58. N.Czech, W.Eber and F.Schmitz, Mater.Sci.Tech., 1986, 2, 244-249

59. K.Y.Kim, S.H.Kim, K.W.Kwan and L.H.Kim, Oxid.Met., 1994, 41, 179-188

60. http://sciencelinks.jp/j-east/article/200604/000020060406A0061342.php

61. J.H.Sun, E.Chang, B.C.Wu and C.H.Tsai, Surf.Coat.Tech., 1993, 58, 93-99

62. R.Knight, D.Zhagxiong, E.H.Kim, R.W. Smith, P.Sahoo and B.Bucci, proceedings of 15th International Thermal Spray Conference, May 1998, NIC, France, vol.II p 1549

63. C.Leyens,K.Fritcher, M.Peters and W.A.Kayser, Surf.Coat.Technol., 1997, 94-95, 155-161

64. W.Beele, N.Czech,W.J.Quadakkers and W.Stamn, 1997, 94-95, 41-48

65. Q.M.Wang, Y.N.Wu, P.L.Ke, H.T.Cao, J.Gang, C.Sun and L.S.Wen, Surf.Coat.Techonol., 2004, 186, 389-397

66. M.P.Bacos, B. Girare, P.Josso and C.Rio, Surf.Coat.Technol., 2003, 162, 248-260

67. M.P.Bacos, P. Josso, N.Vialas, D. Poquillon, B.Pieraggi, D.Monceau, J.R.Nicholls, N. Simms, A.Encinas-Oropesa, T.Ericsson and S.Stekovic, Appl.Thermal.Eng., 2004, 24, 1745-1753

68. F.Wu, H. Murakami, A. Suzuki and H. Harada, Surf. Coat. Tech., 2003, 168, 62-69

69. P. Kuppusami, H. Murakami and T. Ohnuma, J. Vac. Sci. Tech., A 2004, 22, 1208-1217

70. Y.N. Wu, A. Yamaguchi, H.Murakami and S. Kuroda, J. Mater. Res., 2007, 22, 206-216

71. A. Suzuki, Y.Wu, A. Yamaguchi, H. Murakami and C.M.F. Rae, Oxid. Met., 2007, 68, 53-64

72. K. Kawagishi, A. Sato and H. Harada, JOM (2008) 31-35 (www.tms.org/jom.html)

73. M. Seraffon, N.J. Simms, J. Sumner, J.R. Nicholls, Surf. Coat Technolo. 2011, 206, 1529-1537

74. K.L.Luthra and J.H.Wood, Thin Solid Films, 1984, 119, 271-280

75. R.Bauer and H. Gunling, Thin Solid Films, 1982, 1, 3-20

76. G.W.Goward, J.Eng.Gas Turbines Power, 1986, 108, 421-425

77. B.A.Pint, Oxid.Met., 1996, 45, 1-37

78. A. Aygun, A.L. Vasiliev, N.P. Padture, Xinqing Ma, Acta. Mat, 2007, 55, 6734-6745

79. M.H. Vidal-Setif, N. Chellah, C. Rio, C. Sanchez, O. Lavigne, Surf. Coat. Tech, 2012, 208, 39-45

80. Y. Wang and G. Sayre, Surf.Coat. Technol, 2009, 203, 2186-2192

81. M. Abbas, L. Guo and H. Guo, Ceram. Intl, 2013, 39, 5103-5111

82. C.V. Cojocaru, D. Lévesque, C. Moreau, R.S. Lima, Surf. Coat. Tech., 2013, 216, 215-223

83. X. Xie, H.Guo, S. Gong, H. Xu, J. Europ.Ceram. Soc, *2011,* 31,*1677-1683*

84. J. M. Drexler, A. Aygun, D. Li, R.Vaßen, T.Steinke and N.P. Padture, Surf. Coat.Technol, 2010, 204, 2683-2688

85. M. S. Majewski, C. Kelley, W. Hassan, W.Brindley, E.H. Jordan, M.W. Renfro, Surf. Coat.Tech, 2011, 205, 4614-4619

86. S.W. Kandebo, Aviation week, 1998, 148, 34-35

87. J.C.Schaffer, W.H.Murphy, W.B.O'Connor, B.J.Nagaraj and H.V.Vakil, US patent 5,538,796 (1996)

88. D. Renusch and M. Schutze, Mater. Sci. Firum, 2008, 595-598, 151-158

89. D. Renusch and M. Schutze, Mater. Corr., 2008, 59, 547-555

90. T. Sadowski, P. Golewski, Comput. Mat.Sci, 2011, 50,1326-1335

91. A.H. Rossette, C.Z.Mazur, A. Demeulenaere, J.A. R.L. Hernández, App.Ther. Eng, 2009, 29, 3056-3065

92. T. Beck, M. Schweda and L. Singheiser, Proc. Eng, 2013, 55,191-198

93. T. Beck, R.Herzog, O.Trunova, M. Offermann, R.W. Steinbrech, L. Singheiser, Surf. Coat.Technol, 2008, 202, 5901-5908

94. M.S. Majewski, C.Kelley, J.Lake, M.W. Renfro, W. Hassan, W. Brindley, E. H. Jordan, Surf.Coat.Technol, 2012, 206, 2751-2758

95. I. Gurrappa and A. Sambasiva Rao, Surf. Coat. Technol., 2006, 201, 3016- 3029

96. Y. Ito, Handbook of Advanced Ceramics (Second Edition*), 2013, 789-806*

97. M.T. Kim and S.W. Lee, *Appl. Thermal Eng., 2012*, 40, *304-310*

98. H. Aleksanoglu, A. Scholz, M. Oechsner, C. Berger, M. Rudolphi, M. Schütze and W. Stamm, *Intl. J. Fatigue, 2013*, 53, *40-48*

99. A.Kulkarni, A. Goland, H. Herman, A. J. Allen, T. Dobbins, F. DeCarlo, J. Ilavsky, G.G. Long, S. Fang and P. Lawton, *Mater. Sci. Eng., A 2006*, 426, *43-52*

100. M.Okazaki, S.Yamagishi, Y. Yamazaki, K. Ogawa, H.Waki and M. Arai, Intl. J.Fatigue, *2013*, 53, *33-39*

101. R.Ghasemi, R. Shoja-Razavi, R. Mozafarinia and H. Jamali, *Ceram. Intl, 2013*, 39, *8805-8813*

102. I. Gurrappa and I.V.S.Yashwanth, Design and Development of Smart Coatings for Gas Turbines, "Gas Turbines" ISBN: 978-953-307-146-6, Injeti Gurrappa (Editor), 2010 SCIYO Publishers, USA, pp 65-78

103. J.R.Nicholls, N.J.Simms, W.Y.Chan and H.E.Evans, Surf.Coat.Technol., 2001, 149, 236-244

104. I.Gurrappa, J. Coat.Tech.Res., 2008, 5, 385-391

105. I.Gurrappa, European Commission Marie Curie Final Report, July 2008

106. K. Schneider, H.G. Von Arnim and H.W. Grunling, Thin Solid Films, 84 (1981) 29-36

107. N. Swindells, Mater. Sci.Tech., 1986, 2, 250-255

108. S.P.Cooper and A. Strang, Mater. Sci.Tech., 1986, 2, 256-261

109. A.Strang, Mater. Sci. Tech., 1986, 2, 243-249

110. N. Czech, W. Esser and F. Schmitz, Mater. Sci. Tech., 1986, 2, 244-249

111. G.W. Goward, Mater. Sci. Tech., 1986, 2, 194-200

INDEX